Pre-Algebra by Design

Second Edition

Russell F. Jacobs

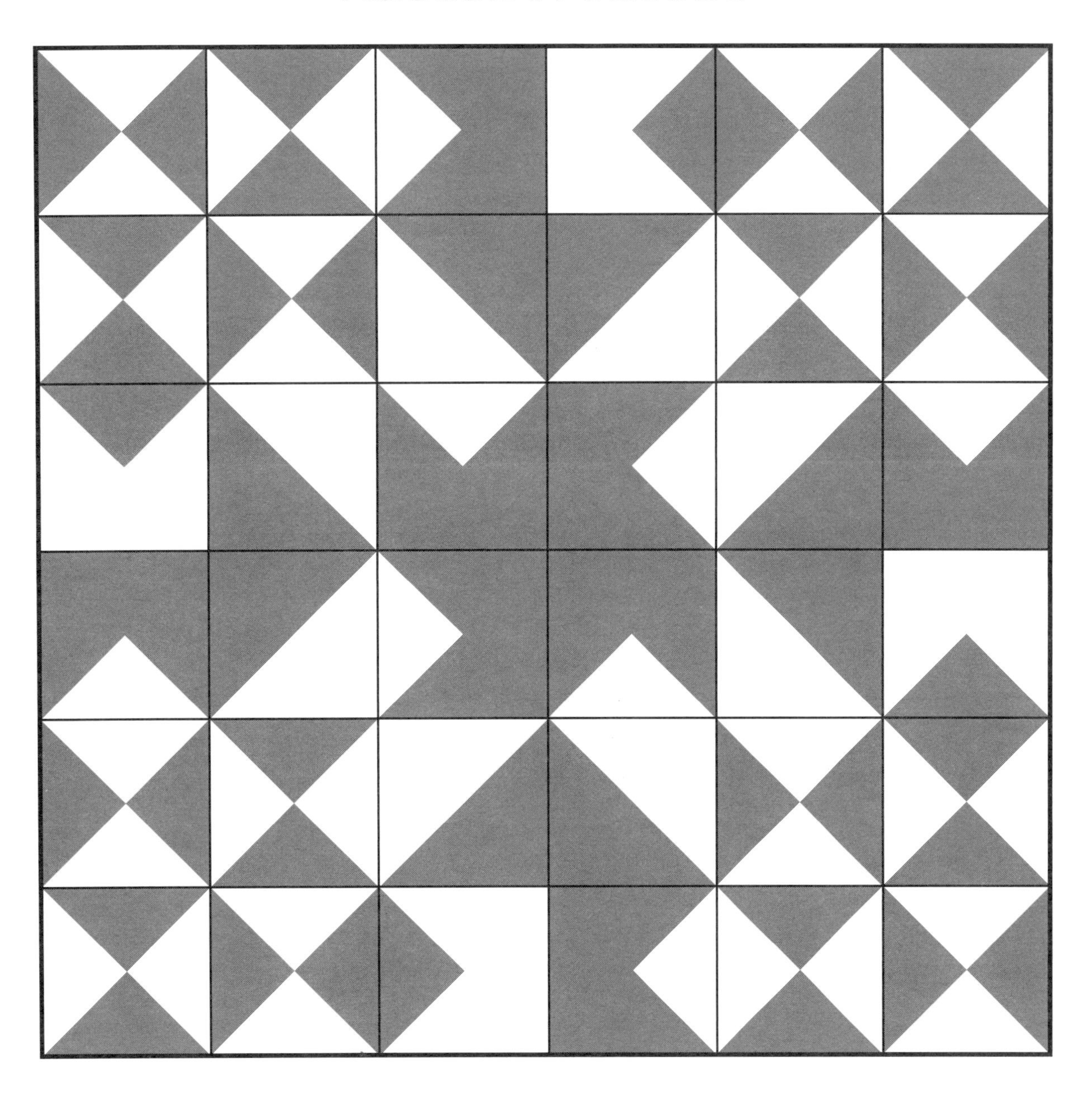

Jacobs Publishing
An imprint of Tessellations

Phoenix, Arizona

AUTHOR **Russell F. Jacobs**

EDITORIAL ASSISTANTS Abbey Naughton, Sonja Jacobs, and Victor Bobbett

The author wishes to credit Patricia Wright for the "search n shade" concept used in **Pre-Algebra by Design**. She introduced this idea in her **Search n Shade** and **Compute a Design** books, published by Jacobs Publishing Company.

To the Teacher

The following directions may be helpful in getting students started with **Pre-Algebra by Design**.

"Work each exercise. Then search for the answer on the grid. Each answer will appear one or more times. Shade each square containing the answer just like the small square next to the exercise is shaded. If the grid is shaded correctly, a pleasing design emerges."

Both the student and the teacher usually can tell at once if the work is not correct. Most of the designs are symmetrical. Any errors in working the exercises or in shading are noticeable. For convenience, an Answer Key for all the activities is included in the back of the book.

Some students will want to create some artistically elegant designs in color. It is recommended that they first do the shading in pencil to make sure the design is correct. Then they can redo the design in color on a separate grid using a color felt pen.

The exercises in **Pre-Algebra by Design** are for review and practice of certain skills and concepts. In exercises involving solution of equations, to avoid guessing of answers, you will probably want students to show the steps of solving each equation. In exercises involving word problems, you will probably want students to write an equation for each problem and show the steps in solving each equation.

Printed in the United States of America

ISBN 978-0-918272-41-6

Table of Contents

ACTIVITY 1 Name ____________________

7	8	16	16	18	6
17	12	1	1	3	9
10	15	20	6	15	4
10	15	3	9	15	4
18	8	1	1	20	6
17	12	5	5	17	12

Simplify.

$2 \times 3 + 4$

$5 + 2 \times 2$

$4 - 3 + 2$

$9 - 6 \div 3$

$15 \div 5 + 2$

$4 \times 3 \div 2$

$16 - 3 \times 4$

$2 + 2 \times 9 - 5$

$20 - 4 \times 3 \div 6$

$36 \div 4 \div 9$

$12 - 3 + 4 \times 2$

$16 - 30 \div 2 + 11$

$6 \times 3 - 2 \times 5$

$12 \div 3 + 3 \times 4$

$24 \div 6 \times 5$

ACTIVITY 2 Name ____________________

36	21	16	7	14	7
35	2	3	35	2	9
16	14	6	15	25	7
15	21	7	12	8	36
3	2	4	9	2	4
15	8	15	12	25	6

Simplify.

$(12 - 4) \times 2$

$4 \times (3 + 6)$

$2 + (8 - 3)$

$15 - (11 - 4)$

$12 - (6 + 3)$

$7 \times (10 \div 5)$

$(6)(4) \div 6$

$(10 - 3)(4 + 1)$

$5(8 - 3)$

$36 \div (9 \times 2)$

$(11 - 5)4 \div 2$

$(6 + 12) \div (7 - 4)$

$27 - (5 + 7)$

$19 - (4 + 12 \div 2)$

$(19 - 4) + 12 \div 2$

ACTIVITY 3 Name ____________________

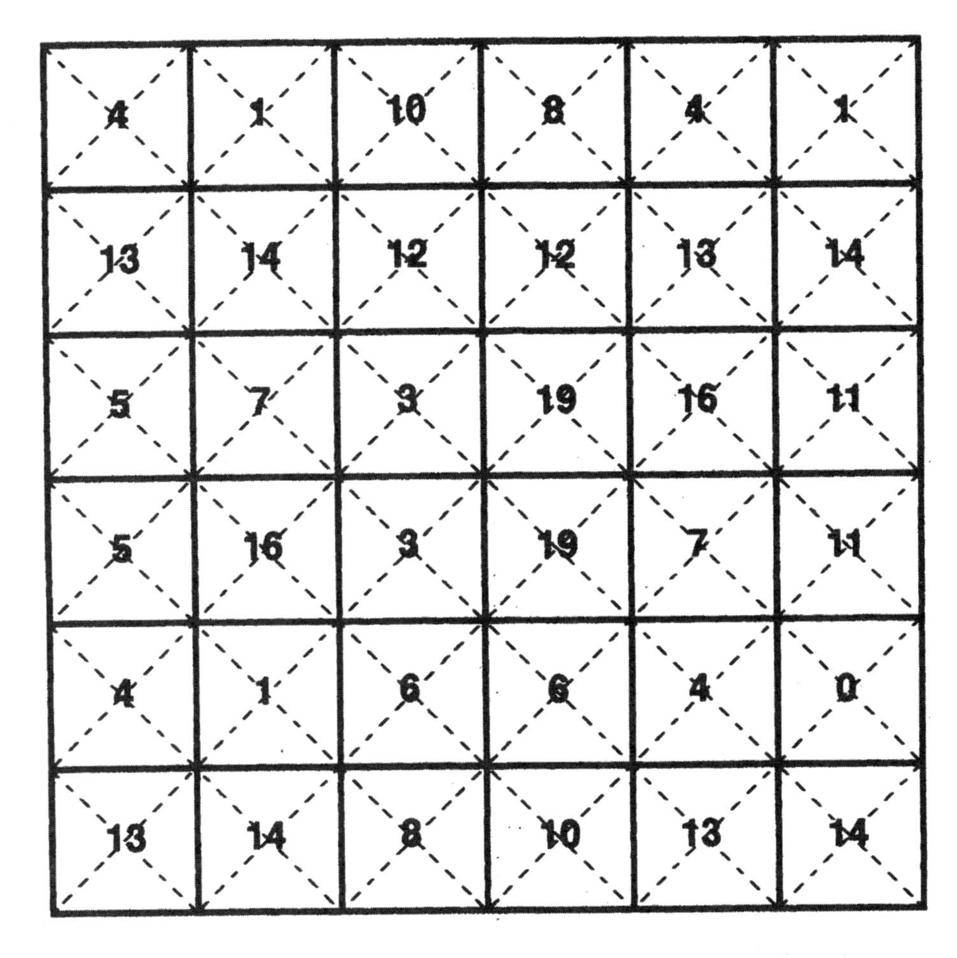

Find each value.

$t + 2 \times 3$ if $t = 4$.

$12 - 2 \times w$ if $w = 3$.

$8 - k + 3$ if $k = 6$.

$y - 10 \div 2$ if $y = 12$.

$3 \times m \div 6$ if $m = 8$.

$16 + n \div 4$ if $n = 12$.

$x - 8 + 7$ if $x = 12$.

$g - 4 \times 2$ if $g = 16$.

$3 \times v + 2$ if $v = 4$.

$15 - r \div 7$ if $r = 14$.

$4 + 4 \times t$ if $t = 2$.

$2 \times h - 12 \div 3$ if $h = 10$.

$4 \times m \div 12$ if $m = 9$.

$18 - 2 \times k$ if $k = 9$.

$16 \div 2 - 7 \times n$ if $n = 1$.

ACTIVITY 4 Name ____________________

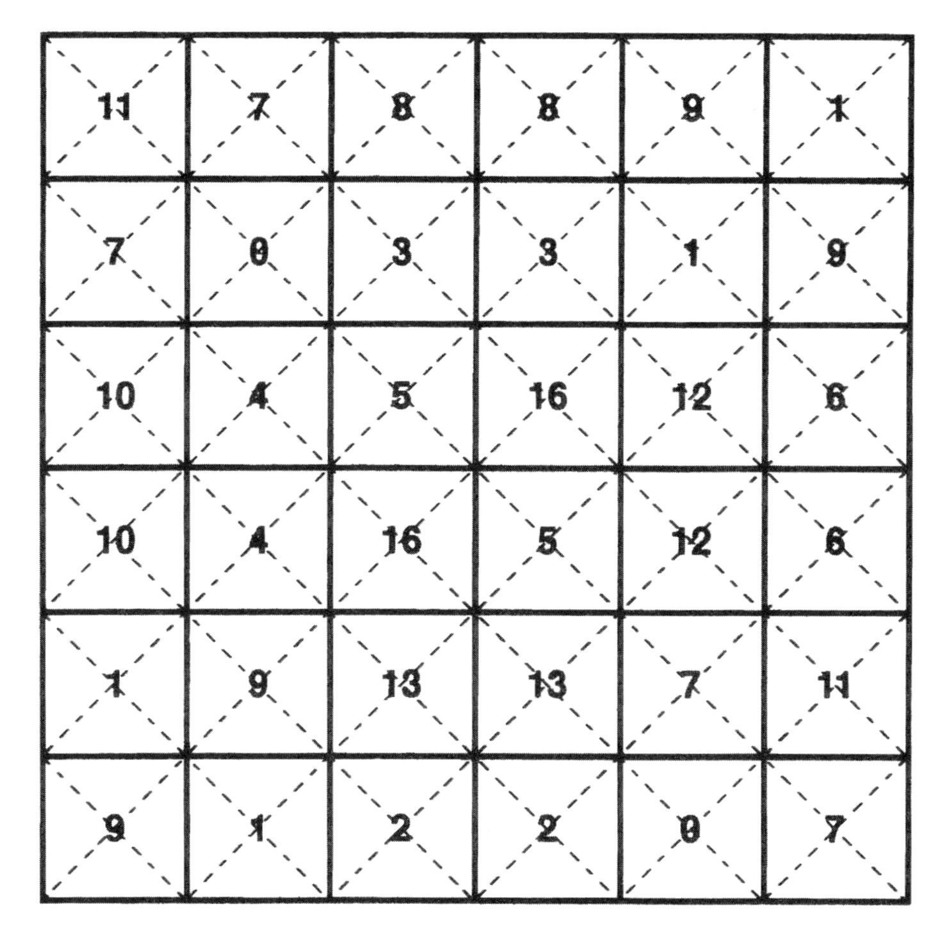

Find each value.

$r + 3 - s$
if $r = 5$ and $s = 6$.

$3 \times m \div n$
if $m = 8$ and $n = 2$.

$a - 3 \times b$
if $a = 20$ and $b = 5$.

$5 \times p + 3 \times q$
if $p = 1$ and $q = 2$.

$v - w \times 3$
if $v = 10$ and $w = 2$.

$c + d \div 5$
if $c = 3$ and $d = 20$.

$x - y + 8$
if $x = 5$ and $y = 3$.

$4 \times m - 5 \times n$
if $m = 4$ and $n = 3$.

$24 - r \times s$
if $r = 3$ and $s = 7$.

$c + 3 \times d$
if $c = 0$ and $d = 3$.

$2 \times k - 4 \times t$
if $k = 8$ and $t = 4$.

$5 + m \div n$
if $m = 12$ and $n = 4$.

$36 - p \times q$
if $p = 3$ and $q = 10$.

$x - y \div 4$
if $x = 21$ and $y = 20$.

$g - h \times 4$
if $g = 37$ and $h = 6$.

ACTIVITY 5

Name ____________________

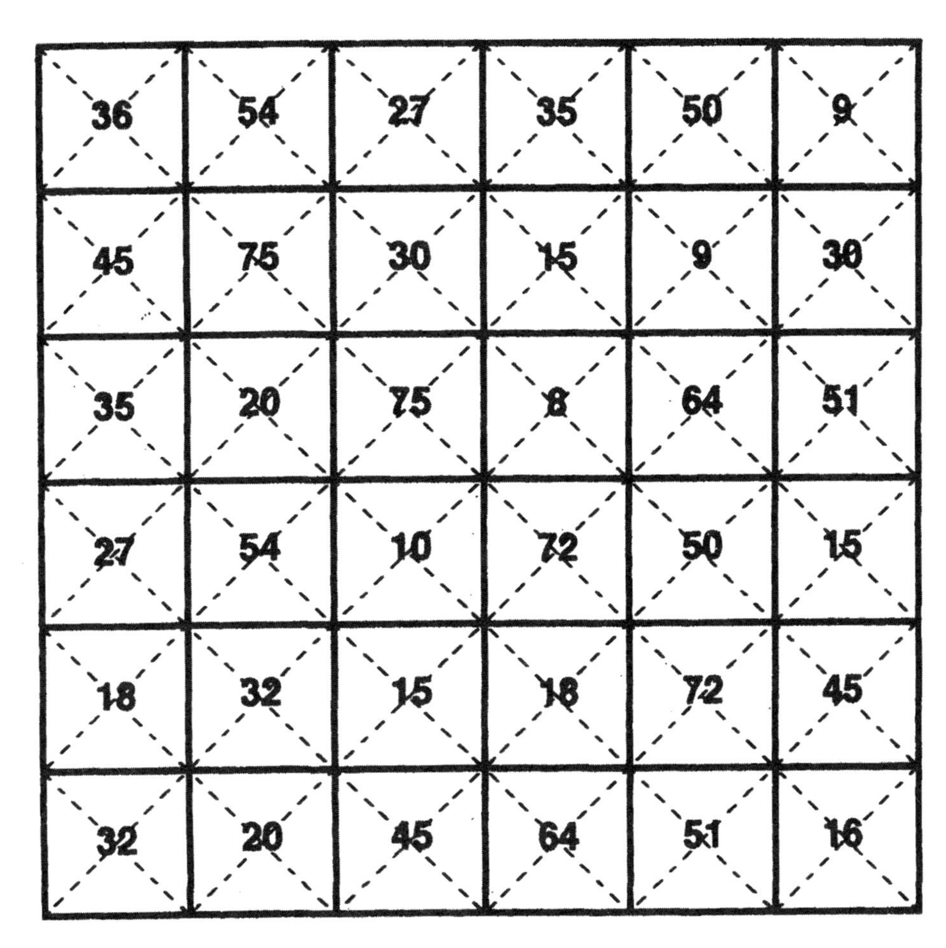

Evaluate.

$A = \ell w$
if $\ell = 5$ and $w = 7$.

$D = rt$
if $r = 25$ and $t = 3$.

$V = \ell wh$
if $\ell = 4$, $w = 6$, $h = 3$.

$A = \frac{1}{2}bh$
if $b = 12$, $h = 9$.

$P = \frac{c}{n}$
if $c = 72$, $n = 8$.

$V = \frac{1}{3}b^2h$
if $b = 4$, $h = 6$.

$H = n(2n - 1)$
if $n = 3$.

$D = \frac{n(n-3)}{2}$
if $n = 9$.

$A = \frac{h(a+b)}{2}$
if $a = 6$, $b = 3$, $h = 4$.

$P = 2(\ell + w)$
if $\ell = 10$, $w = 5$.

$F = \frac{9}{5}c + 32$
if $c = 10$.

$S = \frac{1}{2}gt^2$
if $g = 32$, $t = 2$.

$A = \frac{1}{2}h(a + b)$
if $h = 10$, $a = 5$, $b = 4$.

$C = \frac{5}{9}(f - 32)$
if $f = 68$.

$S = \frac{n}{2}(a + \ell)$
if $n = 6$, $a = 5$, $\ell = 12$.

ACTIVITY 6 Name ____________________

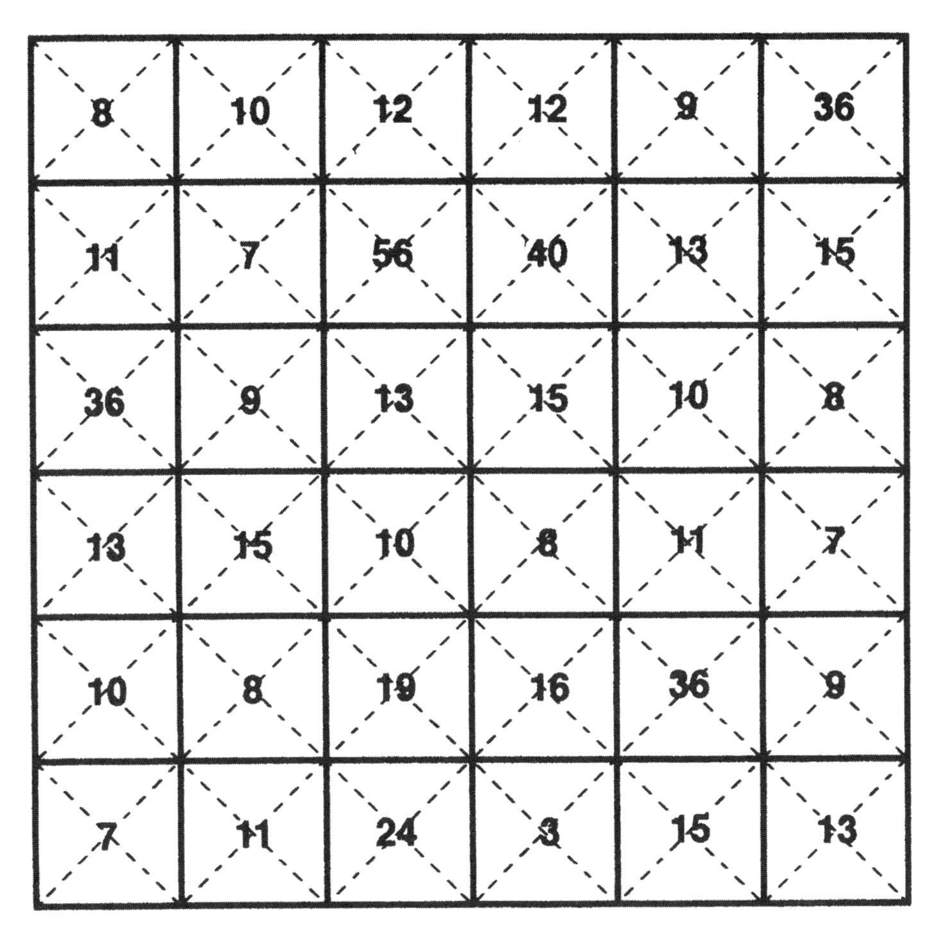

8	10	12	12	9	36
11	7	56	40	13	15
36	9	13	15	10	8
13	15	10	8	11	7
10	8	19	16	36	9
7	11	24	3	15	13

Evaluate.

$A = np$
if $n = 8$ and $p = 7$.

$P = s - c$
if $s = 13$ and $c = 6$.

$R = d \div t$
if $d = 72$ and $t = 6$.

$P = 2\ell + 2w$
if $\ell = 5$ and $w = 3$.

$I = 12f$
if $f = 3$.

$M = 3t - 1$
if $t = 4$.

$A = \frac{1}{2} cd$
if $c = 5$ and $d = 4$.

$N = rc$
if $r = 6$ and $c = 4$.

$T = 10d + 5n$
if $d = 3$ and $n = 2$.

$T = \frac{d}{r}$
if $d = 56$ and $r = 7$.

$S = \frac{n(n-1)}{2}$
if $n = 6$.

$D = dq + r$
if $d = 2$, $q = 4$,
and $r = 1$.

$P = 2^n - 1$
if $n = 2$.

$Y = 3x - 5$
if $x = 6$.

$Y = 2x^2 + 1$
if $x = 3$.

ACTIVITY 7 Name ______________________________

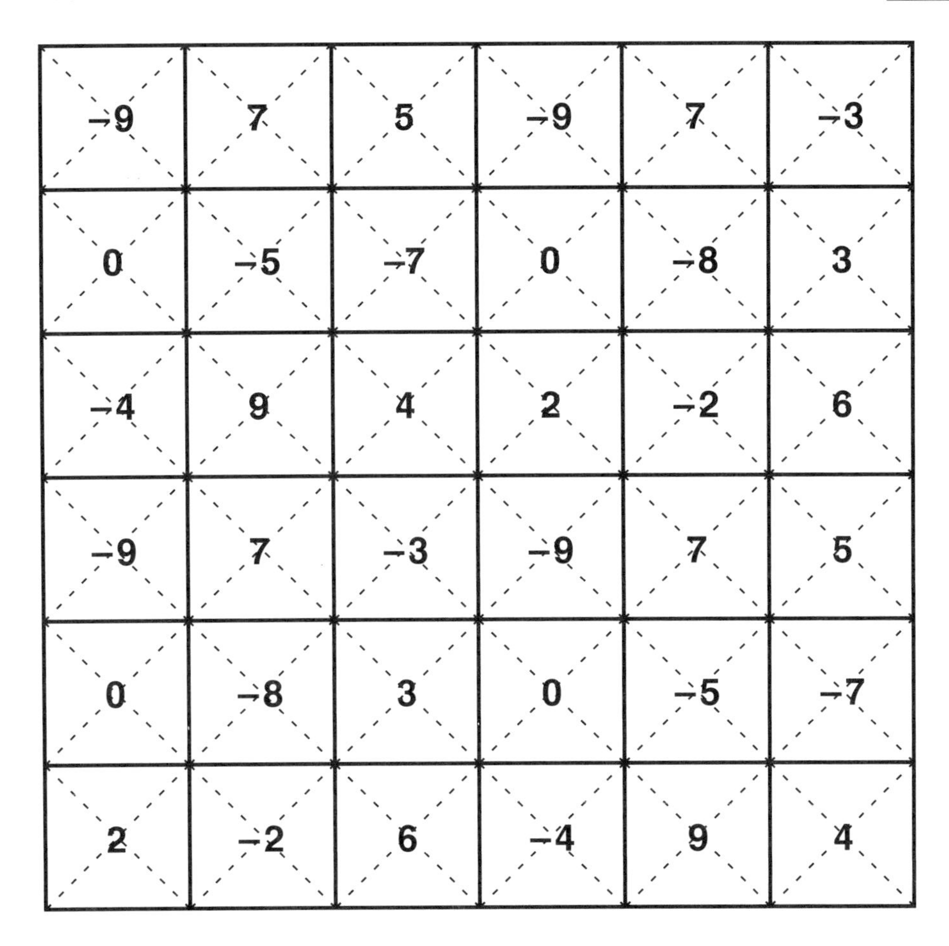

−9	7	5	−9	7	−3
0	−5	−7	0	−8	3
−4	9	4	2	−2	6
−9	7	−3	−9	7	5
0	−8	3	0	−5	−7
2	−2	6	−4	9	4

Find each value.

 One more than −5

 Two less than 0

 Four more than −1

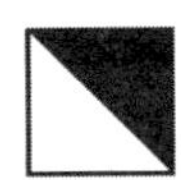 Five less than the opposite of −2

 One-half the opposite of −4

 Two more than −7

 The opposite of 3 less 4

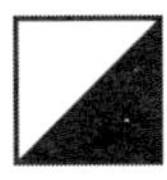 The opposite of the opposite of 4

 One-third of the opposite of −15

 The product of 6 and the opposite of −1

 Ten fewer than the opposite of −19

 21 divided by the opposite of −3

 Ten less than the opposite of the opposite of 2

 Three more than the opposite of 12

 Zero divided by the opposite of −2

ACTIVITY 8

Name ______________________

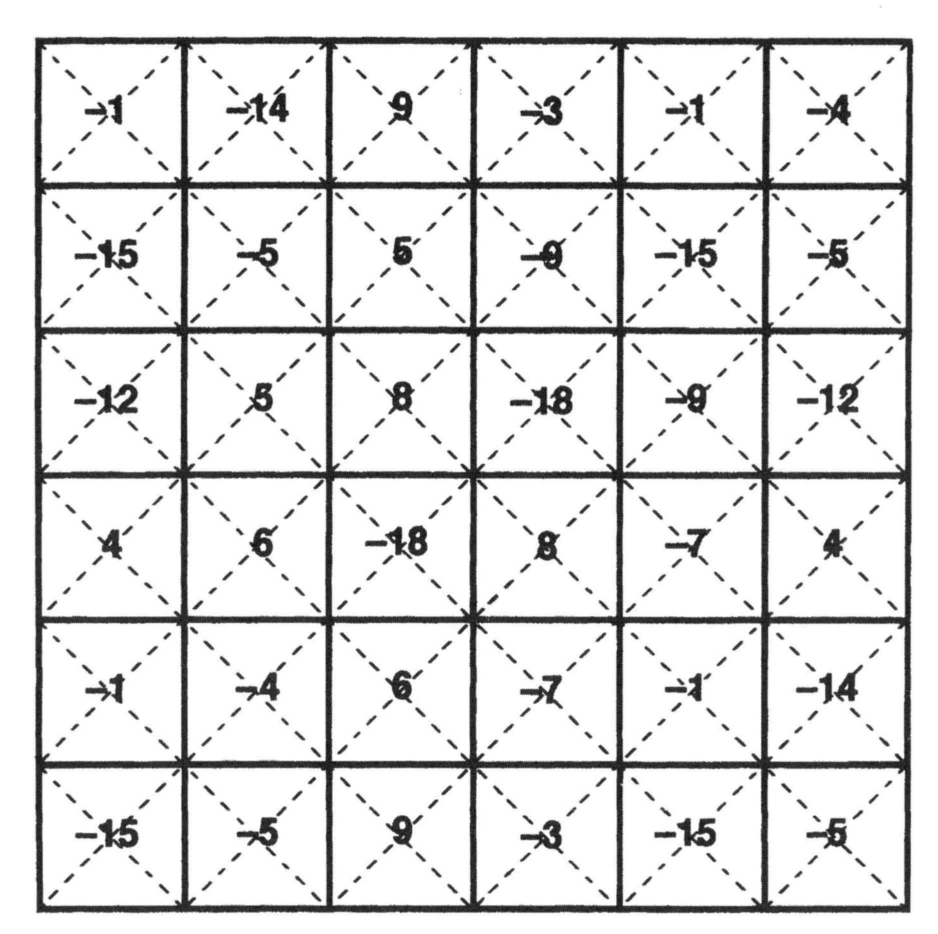

Add.

$8 + (-3)$

$(-4) + (-3)$

$15 + (-9)$

$(-12) + 9$

$(-7) + (-5)$

$(-6) + 15$

$(-6) + 5$

$6 + (-11)$

$(-9) + (-5)$

$(-13) + 9$

$(-3) + 11$

$(-9) + (-9)$

$(-5) + (-4)$

$9 + (-5)$

$(-18) + 3$

ACTIVITY 9

Name ____________________

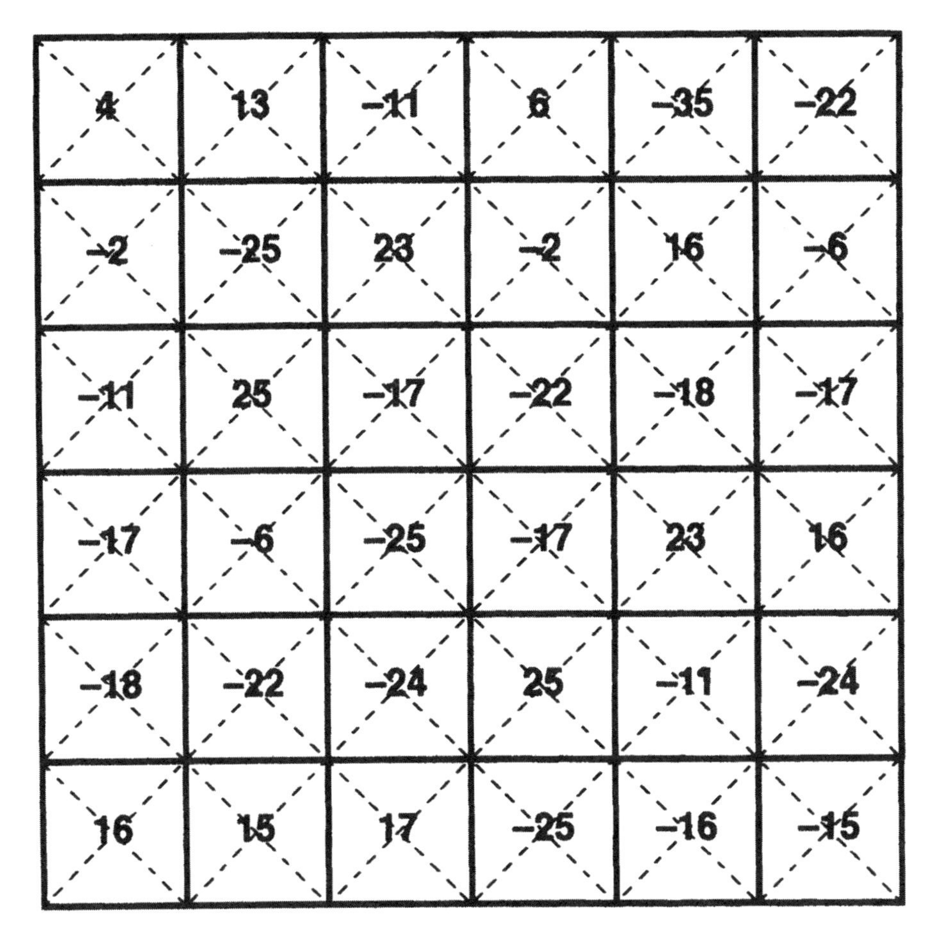

Add.

14 + (−16)

(−23) + 17

(−9) + 22

(−12) + (−13)

8 + (−25)

27 + (−12)

(−37) + 19

(−9) + (−15)

26 + (−37)

29 + (−13)

(−14) + 39

(−10) + (−12)

45 + (−22)

(−19) + (−16)

(−28) + 12

ACTIVITY 10 Name ____________________

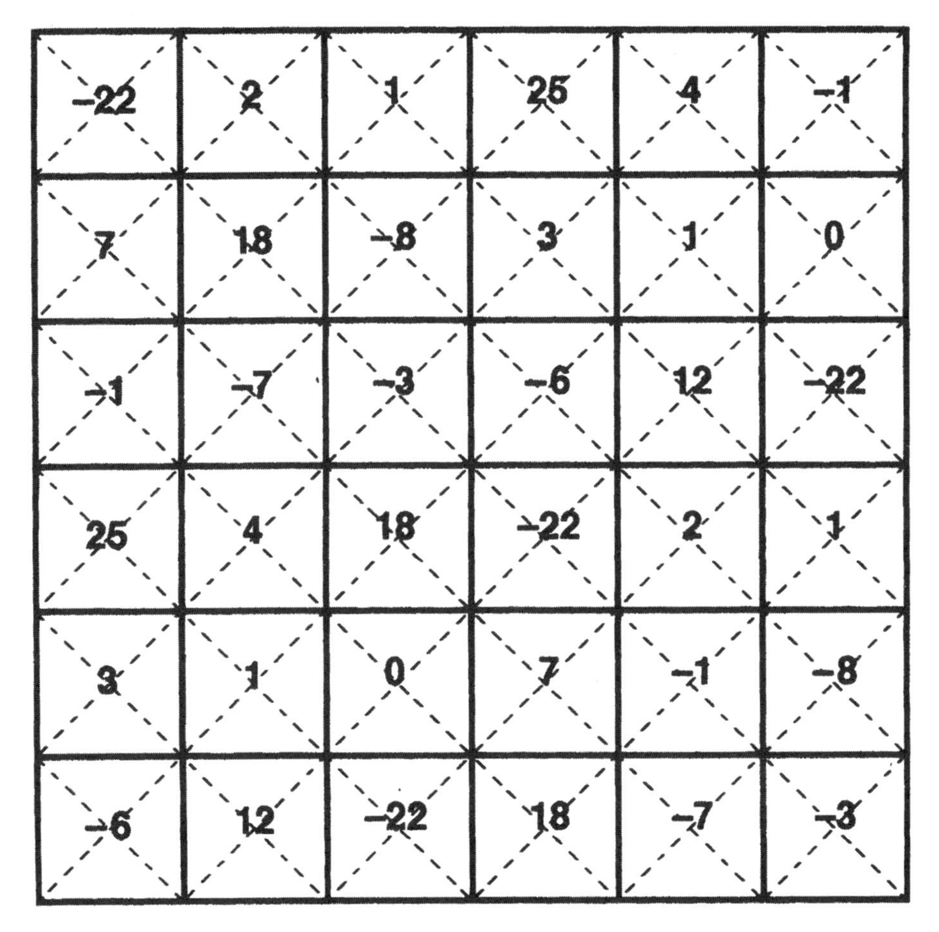

Add.

(−10) + 12 + (−3)

16 + (−9) + 5

24 + (−8) + (−9)

(−13) + 18 + (−12)

25 + 9 + (−16)

29 + (−33) + 7

(−15) + (−6) + 22

35 + (−44) + 11

(−27) + 8 + (−3)

48 + (−56) +12

(−36) + (−4) + 34

29 + 13 + (−50)

72 + (−24) + (−48)

33 + (−17) + 9

(−52) + (−13) + 62

ACTIVITY 11 Name ______________________________

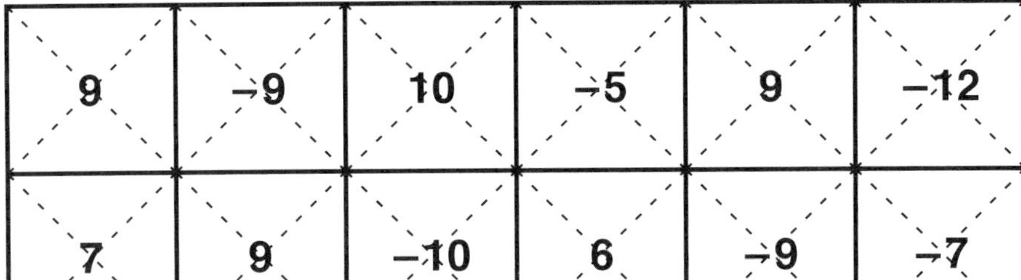

9	−9	10	−5	9	−12
7	9	−10	6	−9	−7
10	5	0	2	−4	−12
−4	−9	−6	13	10	−10
9	7	−12	10	−7	−5
−4	−10	6	−7	−4	−10

Subtract.

$a - b = a + (-b)$
To subtract b from a, add the opposite of b to a.

Example: $5 - (-9) = 5 + 9$
To subtract −9, add its opposite.

$(8) - (-2)$

$(3) - (10)$

$(-4) - (-6)$

$8 - 13$

$0 - (-5)$

$3 - (-4)$

$(-7) - 2$

$2 - (-7)$

$(-4) - (-10)$

$(-12) - (-2)$

$9 - (-4)$

$(-10) - 2$

$(-5) - (-5)$

$(-9) - (-5)$

$5 - 11$

ACTIVITY 12

Name ____________________

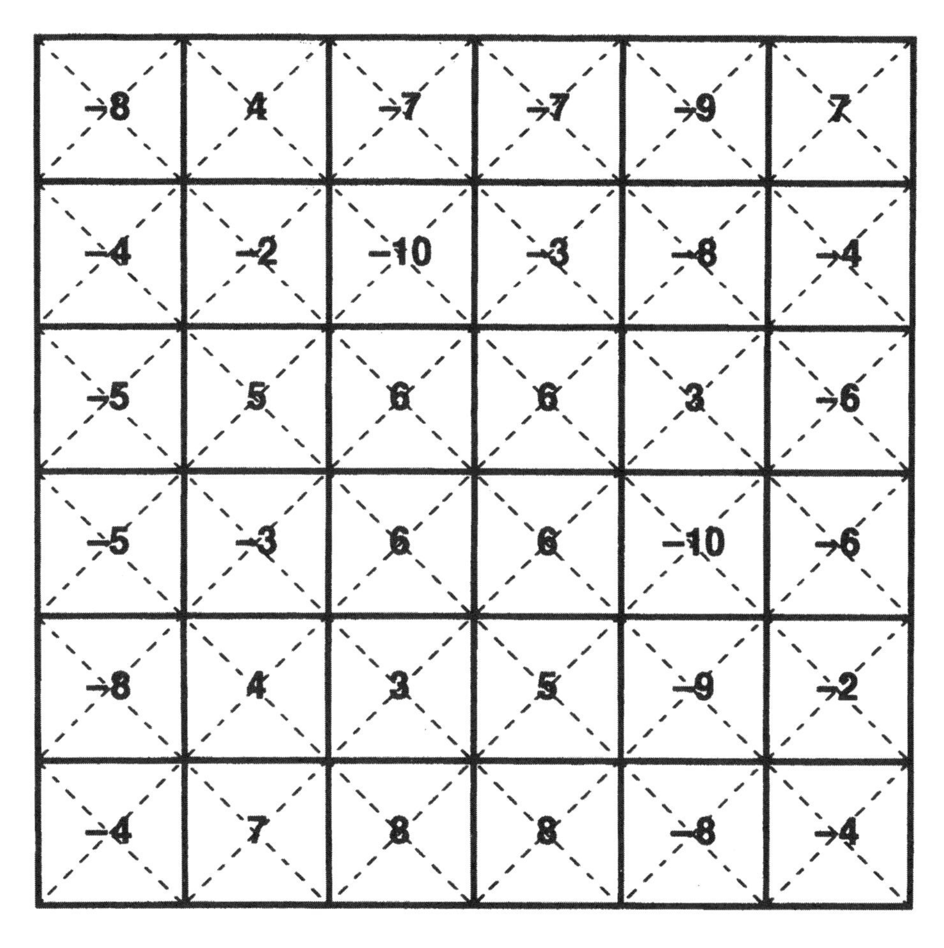

Add or Subtract.

$(-5) - (-2)$

$4 + (-9)$

$(-1) - (-7)$

$10 - 12$

$(-4) - (-7)$

$12 + (-8)$

$(-4) + (-2)$

$15 - 19$

$18 + (-13)$

$(-22) - (-14)$

$0 - 9$

$(-19) + 26$

$(-5) + (-5)$

$25 + (-32)$

$(-12) - (-20)$

ACTIVITY 13 Name ____________________

14	−6	3	7	−14	9
1	5	−14	−6	−5	−1
7	12	0	4	2	−7
1	−14	9	14	−6	3
7	−5	2	12	5	−1
4	12	−7	1	2	0

Add and subtract.

$5 + 2 - 8$

$(-4) + 7 - (-1)$

$6 - 9 + (-2)$

$(-1) - 5 + 8$

$8 + (-3) - (-2)$

$9 + (-3) - 3$

$15 + (-4) - 11$

$(-12) - (-9) + 8$

$(-3) + 11 - 7$

$22 - 18 + (-10)$

$(-8) + 12 - 11$

$19 - (-3) - 13$

$(-24) + 12 - 2$

$35 - 19 + (-4)$

$15 + (-20) - (-19)$

ACTIVITY 14

Name ____________________

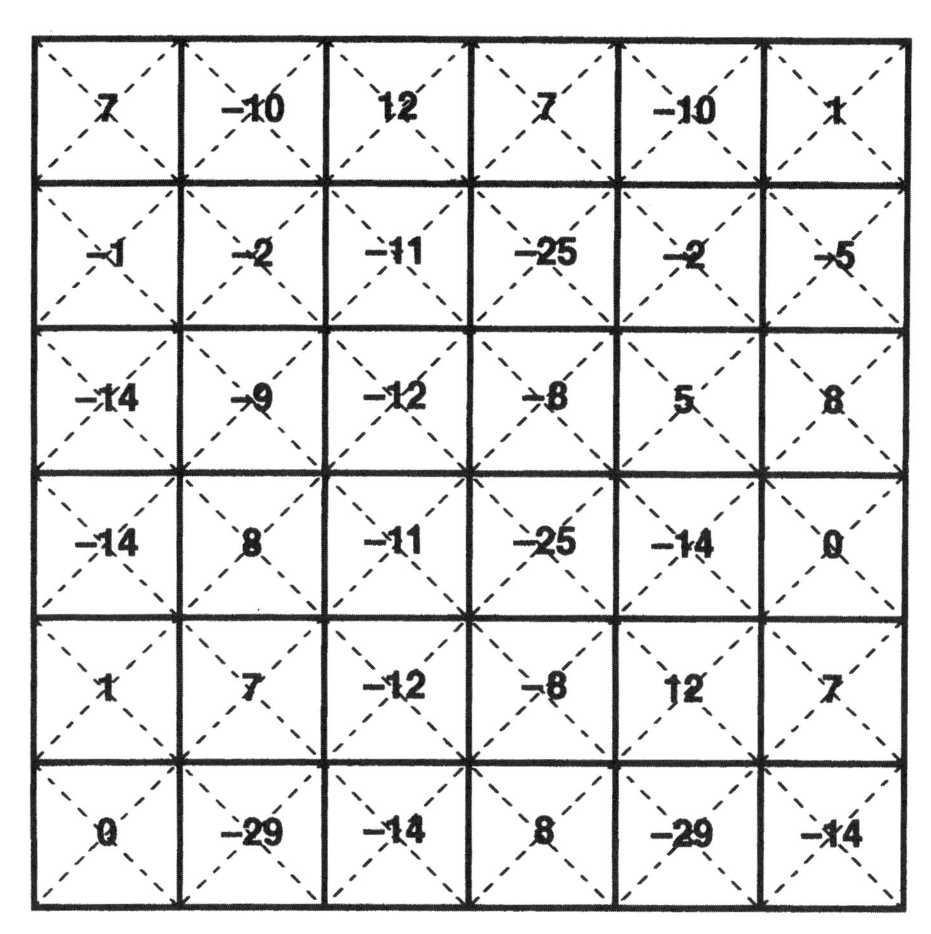

Add and subtract.

$(-12) + 5 - (-2)$

$10 - 18 + (-3)$

$24 + (-8) - 4$

$(-36) + (-4) - (-48)$

$19 - 25 + 4$

$0 - (-12) - 5$

$(-15) + (-8) - 2$

$28 - (-3) - 45$

$(-1) - (-1) + (-1)$

$(-29) + 15 - (-14)$

$20 - 40 + 10$

$(-18) - 5 + (-6)$

$34 - 45 + 12$

$(-56) + 36 - (-12)$

$6 - 9 + 3 - 12$

ACTIVITY 15

Name ______________________

36	−18	−48	−18	−48	36
−12	−24	−10	−10	42	12
0	10	−54	−54	−7	−36
12	−56	−54	−54	−42	−12
−36	56	24	24	7	0
36	18	−48	−18	−48	36

Multiply.

$(-3) \times 4$

$5 \times (-2)$

$(-6) \times (-3)$

$8 \times (-3)$

$(-4) \times 9$

$(-1) \times (-7)$

$(-3) \times (-12)$

$(-4) \times (-3)$

$(-4) \times (-6)$

$(-9) \times 2$

$(-7) \times (-8)$

$(-9) \times 6$

$6 \times (-8)$

$(-7) \times (-6)$

$(-15) \times 0$

ACTIVITY 16

Name ______________________

−12	−24	−8	−8	−36	−5
48	12	0	5	16	48
6	−16	0	60	5	56
6	−18	16	−20	12	56
48	60	−18	−20	−16	48
−6	−36	−1	−1	−24	24

Multiply.

 (2)(−3)(−2)

(−4)(−1)(−4)

(5)(−1)(−1)

(−2)(3)(−1)

(4)(−2)(−2)

(6)(2)(−3)

(−8)(2)(−3)

(5)(−2)(2)

(−2)(−2)(−2)

(−1)(−1)(−1)

(3)(−3)(2)

(−3)(5)(−4)

(−5)(0)(−2)

(−6)(−2)(−2)

(4)(−7)(−2)

Name ______________________

-27	18	54	-36	-45	-54
-40	120	56	-40	54	16
48	40	-45	18	42	40
56	18	-40	16	-45	-40
40	48	54	-36	40	42
64	120	-45	18	54	-48

Multiply.

$(-2)(-5)4$

$8(-3)(2)$

$(-4)(7)(-2)$

$(-3)(-3)(-3)$

$6(-1)(-9)$

$(-3)(-4)(-3)$

$(-5)(-4)(-2)$

$4(-4)(-4)$

$(-6)(2)(-4)$

$(-9)(-2)(-3)$

$(-1)(3)(-6)$

$(-2)(-2)(-2)(-2)$

$7(-2)(-3)$

$(-3)(-3)(-5)$

$(-2)(3)(-4)(5)$

ACTIVITY 18

Name ____________________

−12	5	−9	−12	7	−9
4	−2	1	−7	−2	−4
9	0	−3	8	0	−6
−7	−8	−4	4	−8	1
8	−2	−6	9	−2	−3
−12	5	−9	−12	7	−9

Divide.

$(-12) \div (-3)$

$(15) \div (-5)$

$(-24) \div 6$

$(-20) \div (-4)$

$(-27) \div 3$

$0 \div (-19)$

$(-14) \div 2$

$(-13) \div (-13)$

$18 \div (-9)$

$(-36) \div (-4)$

$(-56) \div (-8)$

$42 \div (-7)$

$(-16) \div 2$

$48 \div (-4)$

$(-32) \div (-4)$

ACTIVITY 19

Name ____________________

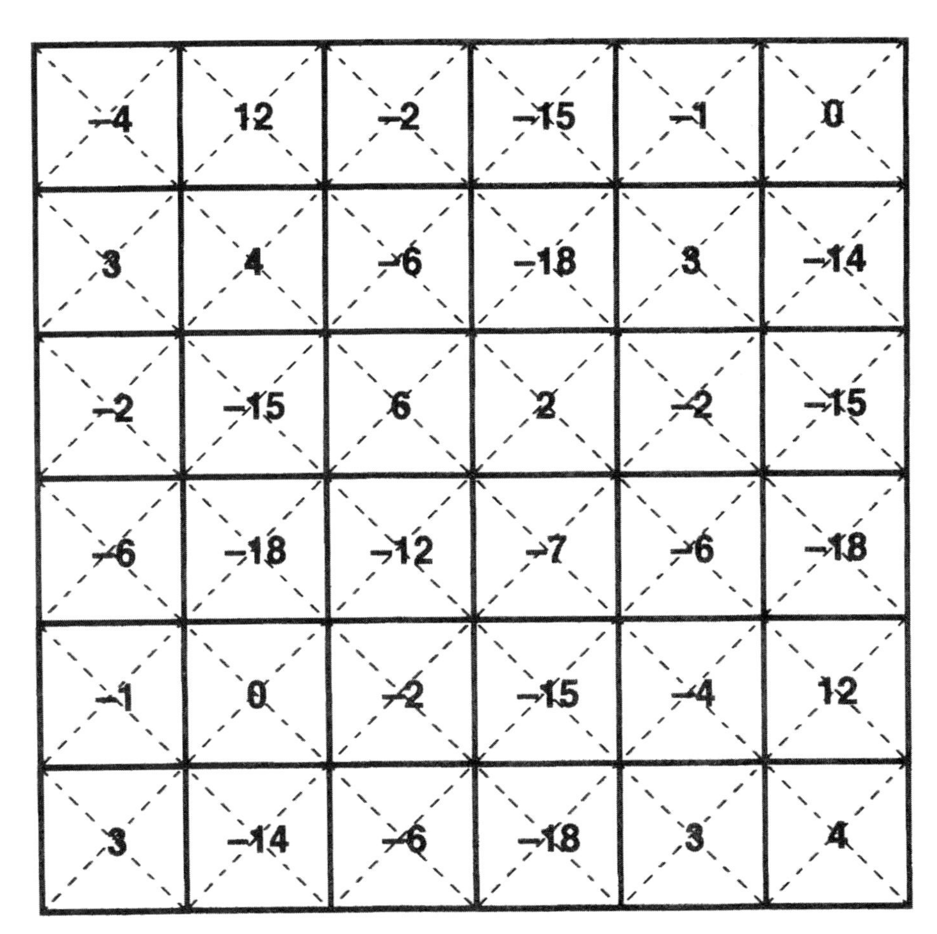

−4	12	−2	−15	−1	0
3	4	−6	−18	3	−14
−2	−15	6	2	−2	−15
−6	−18	−12	−7	−6	−18
−1	0	−2	−15	−4	12
3	−14	−6	−18	3	4

Multiply and divide.

$(-3) \times 4 \div (-2)$

$36 \div (-12) \times 2$

$(-8) \times (-3) \div 6$

$(-42) \div 7 \times (-2)$

$(-8) \div (-4) \times (-2)$

$6 \times 3 \div (-9)$

$9 \times (-4) \div (-12)$

$7 \times (-8) \div 4$

$(-63) \div 7 \times 2$

$(-24) \div 4 \times 2$

$(-7) \times (-7) \div (-7)$

$(-6) \times 0 \div (-2)$

$3 \times 2 \div (-6)$

$(-6) \times (-10) \div (-4)$

$(-4) \times (-8) \div 16$

Name ______________________

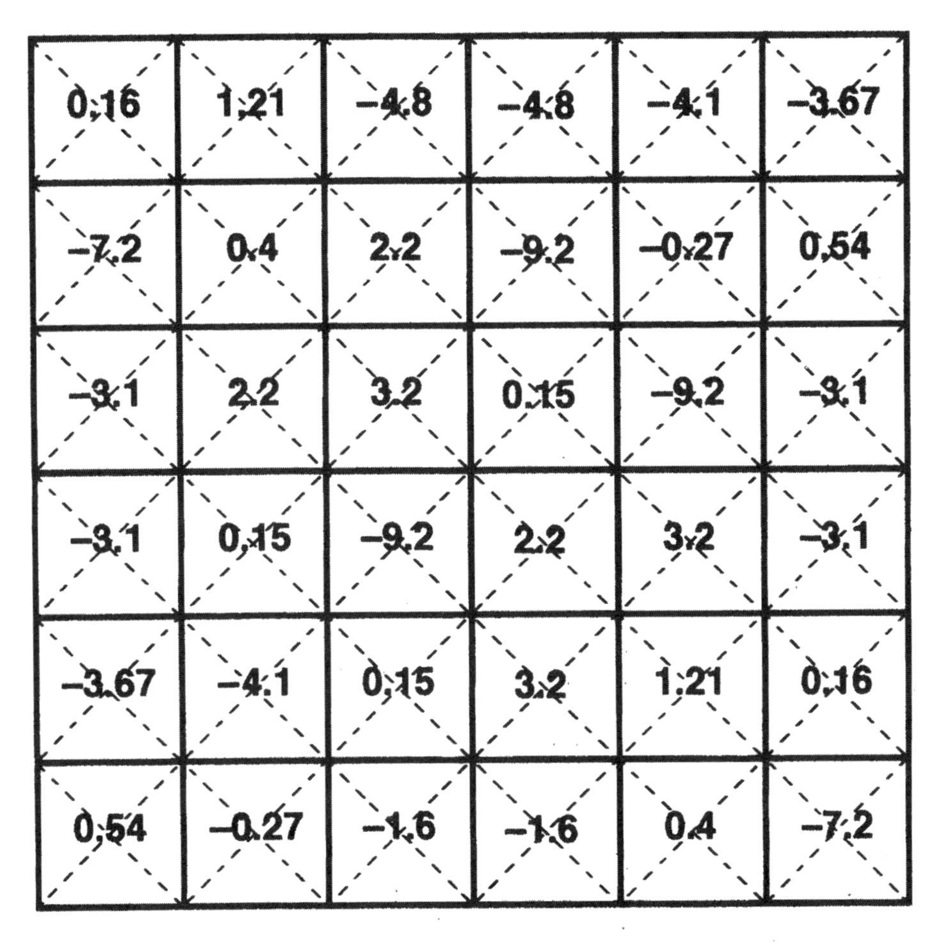

0.16	1.21	−4.8	−4.8	−4.1	−3.67
−7.2	0.4	2.2	−9.2	−0.27	0.54
−3.1	2.2	3.2	0.15	−9.2	−3.1
−3.1	0.15	−9.2	2.2	3.2	−3.1
−3.67	−4.1	0.15	3.2	1.21	0.16
0.54	−0.27	−1.6	−1.6	0.4	−7.2

Add or subtract.

(4.5) + (−2.3)

(−5.8) − (−1.7)

(−0.08) + (−0.19)

(0.8) − (3.9)

(−5.5) − (−8.7)

(−9.1) + (7.5)

(−18.1) + (10.9)

(−0.25) + (0.41)

(−0.03) − (−0.18)

(7.5) − (12.3)

(−6.0) + (−3.2)

(−0.08) + (1.29)

(−6.9) − (−7.3)

(0.45) − (−0.09)

(−3.8) + (0.13)

Name ____________________

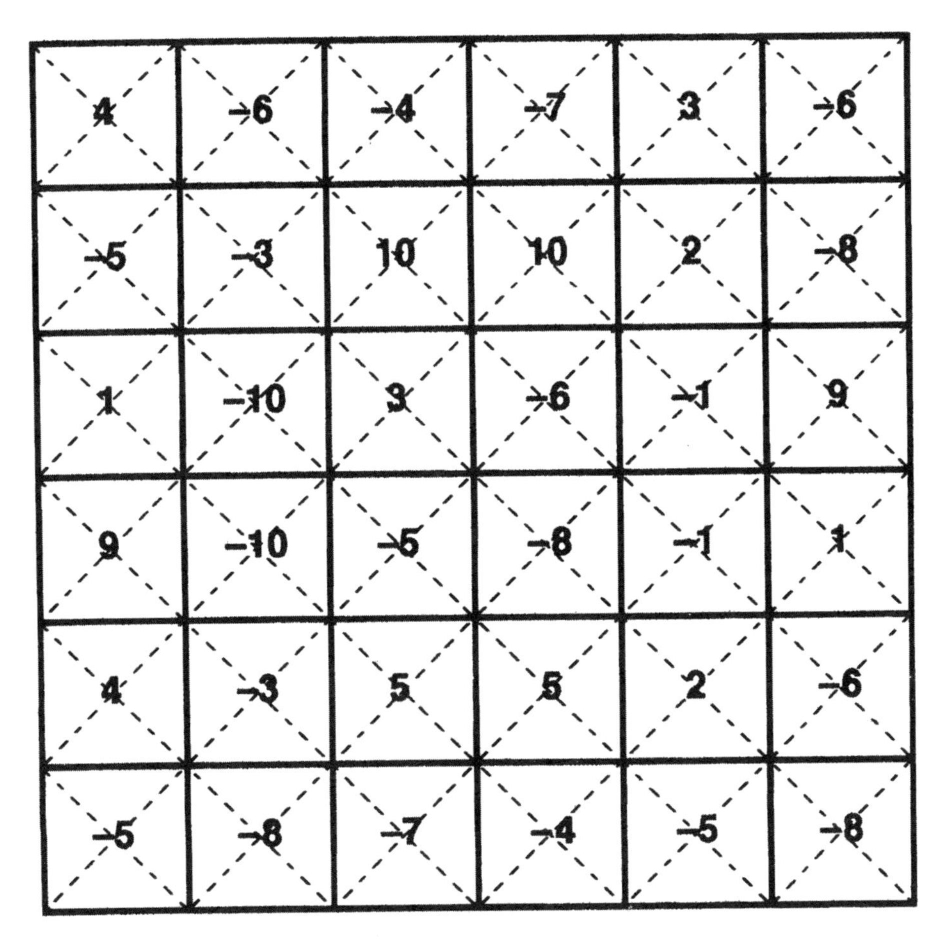

Solve for y.

$y = -2x + 5$ if $x = 4$.

$y = -x + 1$ if $x = -3$.

$y = 3x - 7$ if $x = 2$.

$y = -3x - 10$ if $x = -2$.

$y = x + 5$ if $x = -10$.

$y = -\frac{1}{2}x + 3$ if $x = -4$.

$y = \frac{1}{3}x - 10$ if $x = 12$.

$y = -6x - 5$ if $x = -1$.

$y = -\frac{1}{4}x - 2$ if $x = -16$.

$y = 0.5x + 1.5$ if $x = 3$.

$y = -\frac{1}{5}x - 5$ if $x = 15$.

$y = -4x + 6$ if $x = -1$.

$y = 15 - 3x$ if $x = 2$.

$y = -x - 6$ if $x = 4$.

$y = -\frac{3}{4}x + 2$ if $x = 12$.

Name ______________________

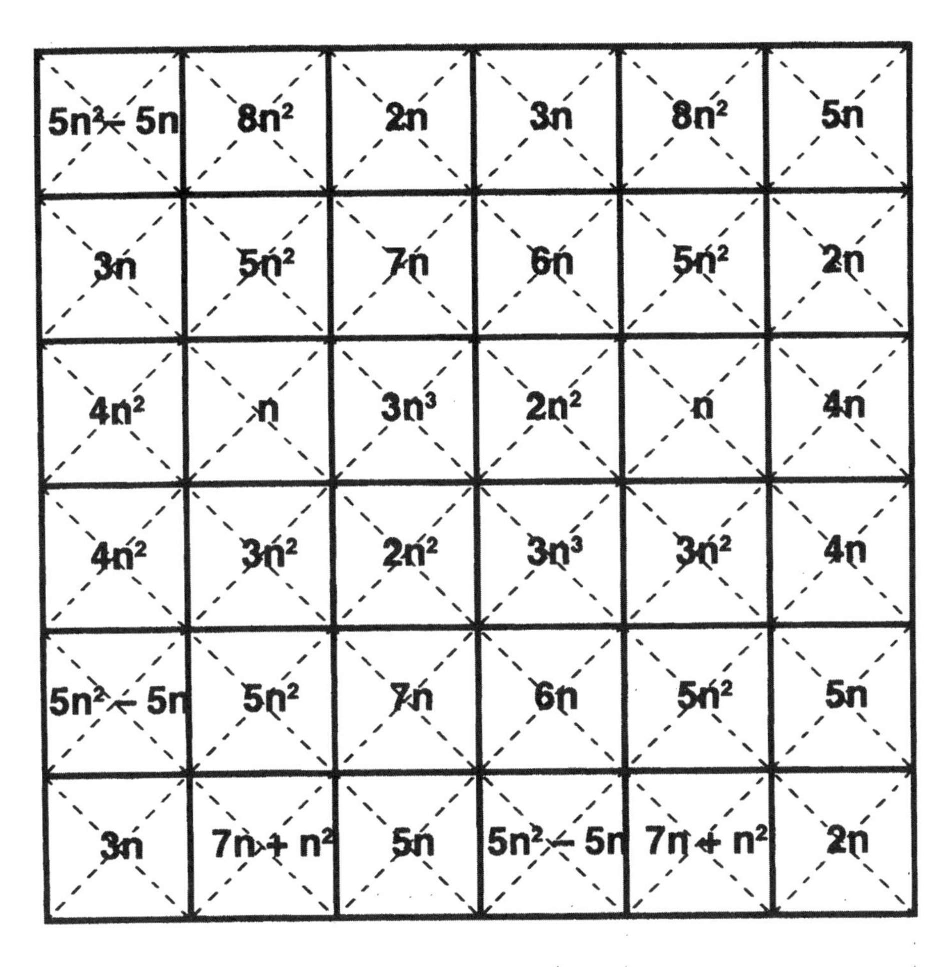

Add and subtract like terms.

 $2n + 3n$

 $10n - 7n$

 $n + n$

$7n - n$

 $6n - (2n + 3n)$

 $6n - 2n + 3n$

$12n - 3n - 5n$

 $3n^2 + n^2$

 $4n + 3n + n^2$

 $13n^2 - (5n^2 + 3n^2)$

 $\frac{3}{4}n^2 + \frac{1}{4}n^2 + n^2$

 $5n^3 - n^3 - n^3$

 $4n^2 + n^2 - 5n$

 $n^2 + n^2 + n^2$

$15n^2 - (3n^2 + 4n^2)$

ACTIVITY 23 Name ______________________________

8n	2n	−9n	−10n	2n	−3n
11n	4n	−8n	−0.7n	−5n	−2.2n
−2.2n	0	2.2n	0.7n	−n	11n
−2.2n	−n	−3n	8n	0	11n
11n	4n	−0.7n	−8n	−5n	−2.2n
0.7n	−10n	2n	2n	−9n	2.2n

Subtract like terms.

Example: $(-n) - (-3n) = -n + 3n$
To subtract $-3n$, add its opposite. Answer: $2n$

 $n - 2n$

 $3n - (-n)$

 $(-4n) - n$

 $10n - (-n)$

 $(-4n) - 5n$

 $6n - (-2n)$

 $n - (-n)$

 $(-5n) - (-5n)$

 $3.7n - 1.5n$

 $0.5n - 1.2n$

 $(-3.5n) - (-1.3n)$

 $(-1.9n) - (-2.6n)$

 $n - 5n - (-n)$

 $(-5n) - 7n - (-2n)$

 $(-3n) - (-n) - 6n$

ACTIVITY 24 Name ____________________

$5n$	$-4n$	$-3n$	$2n^3 + n$	$-6n^2$	$7n$
$-3n$	$n^2 + 2n$	$2n^3 + n$	$4n$	$2n^2 - n$	$n^3 - 5n^2$
$-4n$	$-7n$	$2n$	$-5n^2$	$5n^2$	$-4n$
$-5n$	$-6n^2$	$7n$	$5n$	$-4n$	$-7n$
$4n$	$2n^2 - n$	$n^3 - 5n^2$	$-3n$	$n^2 + 2n$	$2n^3 + n$
$-5n^2$	$5n^2$	$-3n$	$2n^3 + n$	$-5n$	$2n$

Add and subtract like terms.

$(-3n) + 5n$

$(-6n) + (-n)$

$n - (-3n)$

$(-3n) + (-n)$

$(-2n) - (-7n)$

$3n^2 - (-2n^2)$

$(-8n) + n - (-2n)$

$(-4n^2) - 2n^2$

$(-n) + 2n - (-6n)$

$3n^2 - n + (-n^2)$

$-n^3 + 3n^3 + n$

$(-3n^2) + (-2n^2)$

$4n - (-n^2) + (-2n)$

$(-n) + (-n) + (-n)$

$n^3 - 2n^2 - 3n^2$

ACTIVITY 25

Name ____________________

$-9x$	0	$15x$	0	$6x$	$9x$
$6x$	$20x$	$12x$	$-10x$	$9x$	$15x$
$8x$	$-8x$	$-9x$	$-6x$	$-24x$	$8x$
$6x$	$10x$	$-6x$	$12x$	$20x$	$6x$
0	$-12x$	$4x$	$20x$	$-8x$	$8x$
$-12x$	$8x$	$15x$	0	$15x$	$12x$

Multiply.

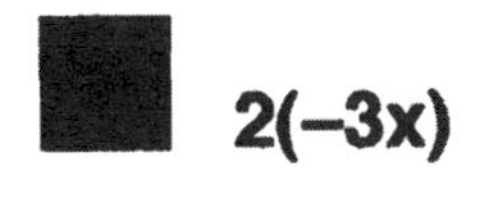

$2(-3x)$

$(-3)3x$

$(-4)(-x)$

$(-2x)5$

$(-6)(-2x)$

$(-2)(-4x)$

$(-2)(-3)(-2x)$

$(-2)(5)(-x)$

$(-3)(2)(-x)$

$(-1)(-1)(-2)(4x)$

$(-1)(3x)(-3)$

$(-x)(0)(-2)$

$(-5x)(-3)$

$(-2)(-3)(-4x)$

$4(-x)(-5)$

ACTIVITY 26 Name ______________________

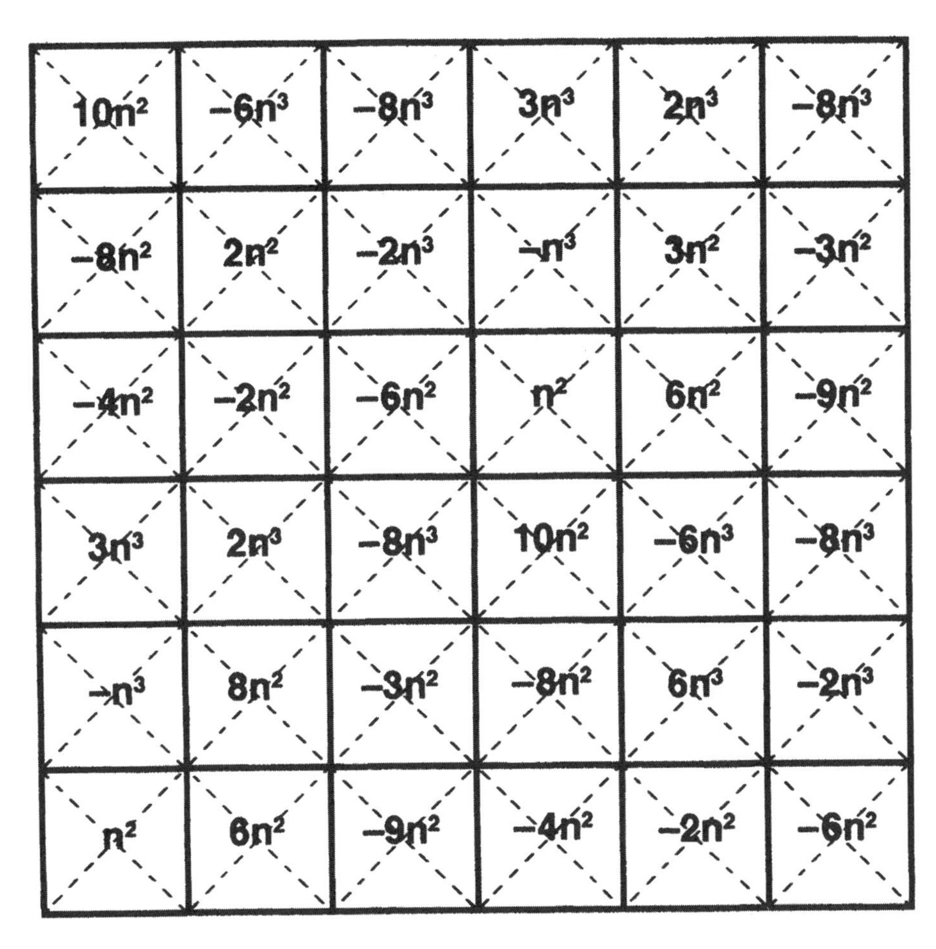

$10n^2$	$-6n^3$	$-8n^3$	$3n^3$	$2n^3$	$-8n^3$
$-8n^2$	$2n^2$	$-2n^3$	$-n^3$	$3n^2$	$-3n^2$
$-4n^2$	$-2n^2$	$-6n^2$	n^2	$6n^2$	$-9n^2$
$3n^3$	$2n^3$	$-8n^3$	$10n^2$	$-6n^3$	$-8n^3$
$-n^3$	$8n^2$	$-3n^2$	$-8n^2$	$6n^3$	$-2n^3$
n^2	$6n^2$	$-9n^2$	$-4n^2$	$-2n^2$	$-6n^2$

Multiply.

$(-3n)(n)$

$(-2n)(-3n)$

$(3n)(-3n)$

$(n)(-n)(2n)$

$(\frac{1}{2}n)(-4n)$

$(-1)(2n)(4n)$

$(-n)(5)(-2n)$

$(-2n)(-n^2)$

$(-1)(-3)(-2n^2)$

$(-n)(-n)$

$(\frac{1}{3}n)(-12n)$

$(-n)(-n)(-n)$

$(-\frac{1}{4}n)(-n)(12n)$

$(-n)(-2n)(-3n)$

$(-2n)(-2n)(-2n)$

ACTIVITY 27

Name ______________________

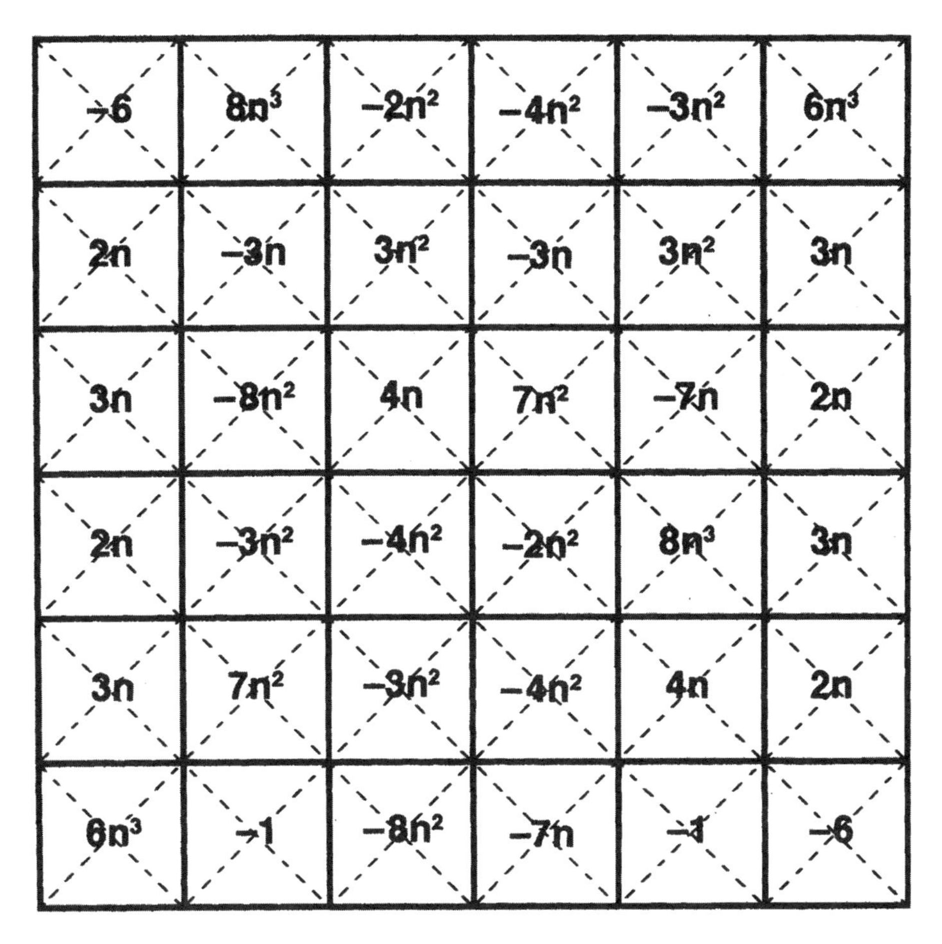

Divide.

$4n^2 \div 2n$

$(-6n^2) \div (2n)$

$8n^3 \div (-n)$

$(-8n^3) \div (-2n^2)$

$12n^2 \div (-2n^2)$

$9n^4 \div 3n^2$

$(-24n^2) \div (-8n)$

$22n^3 \div (-11n)$

$(-16n^4) \div (-2n)$

$(-5n^2) \div 5n^2$

$14n^2 \div (-2n)$

$36n^4 \div (-9n^2)$

$(-56n^3) \div (-8n)$

$(-24n^4) \div (-4n)$

$18n^3 \div (-6n)$

ACTIVITY 28 Name ______________________

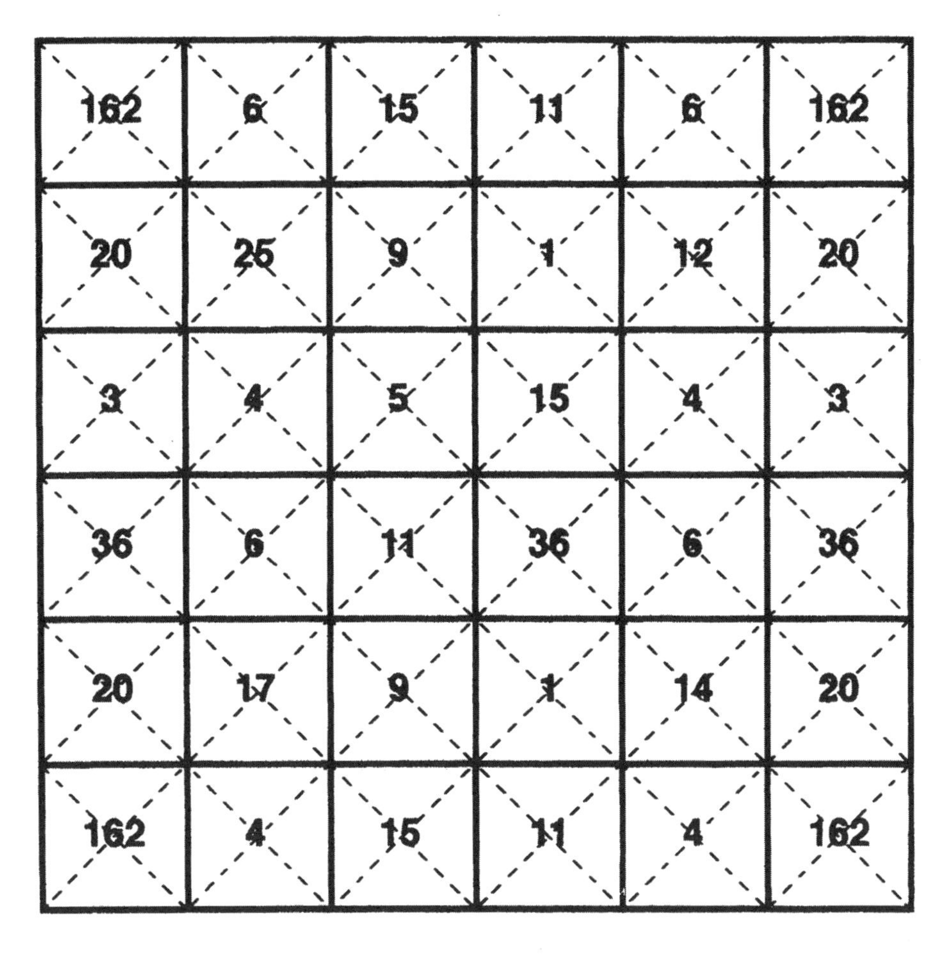

Find the next term in each sequence.

 1, 3, 5, 7, ?

 15, 12, 9, 6, ?

 0, 2, 5, 9, ?

 0, 1, 3, 7, ?

 1, 4, 9, 16, ?

 18, 15, 12, 9, ?

 0, 5, 10, 15, ?

 22, 27, 31, 34, ?

 35, 33, 29, 21, ?

 20, 16, 12, 8, ?

 1, 2, 5, 10, ?

 35, 29, 23, 17, ?

 2, 3, 5, 8, ?

 81, 27, 9, 3, ?

 2, 6, 18, 54, ?

ACTIVITY 29 Name ____________________

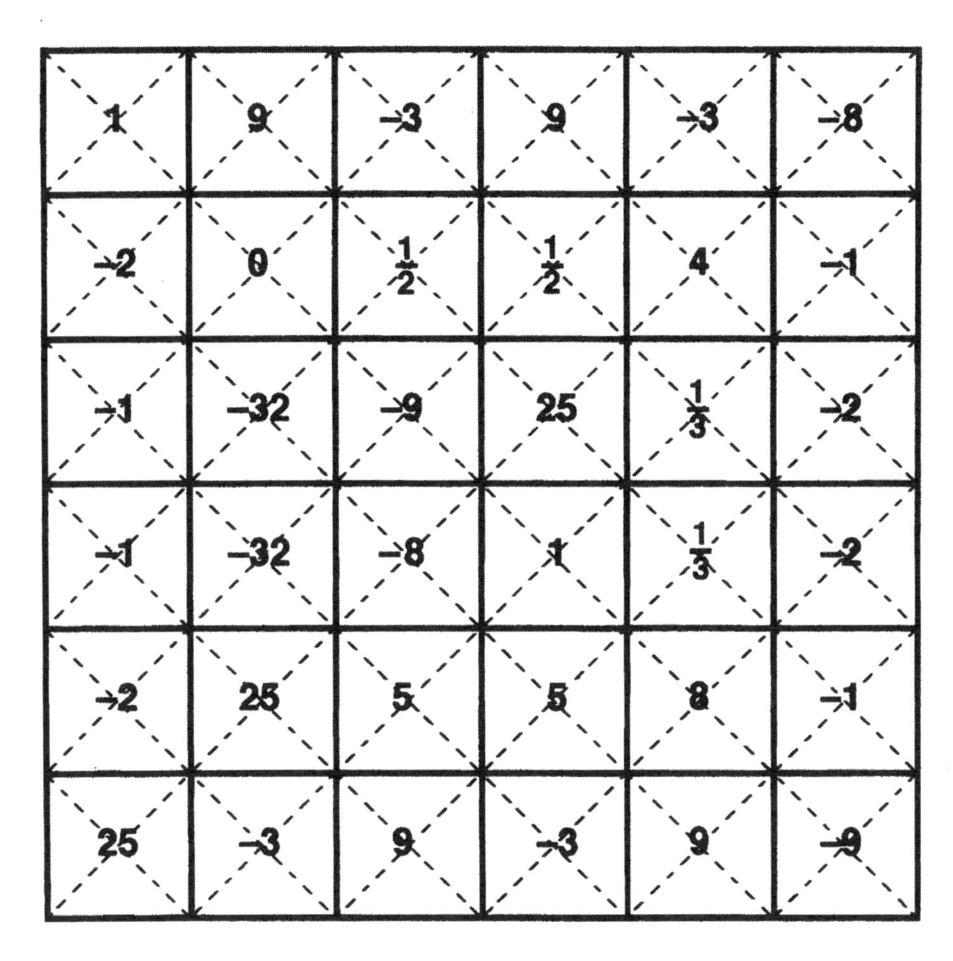

Find the next term in each sequence.

 0, 2, 4, 6, ?

−11, −9, −7, −5, ?

−2, 4, −8, 16, ?

8, 4, 0, −4, ?

 −7, −5, −3, −1, ?

 3, 4, 6, ?

 5, −5, 5, −5, ?

 −14, −8, −4, −2, ?

 −4, −2, 0, 2, ?

 1, 4, 9, 16, ?

 −1, −3, −5, −7, ?

 8, −4, 2, −1, ?

 $-10, -7\frac{1}{2}, -5, -2\frac{1}{2}$, ?

 27, 9, 3, 1, ?

 $4, 2\frac{1}{2}, 1, -\frac{1}{2}$, ?

ACTIVITY 30 Name ____________________

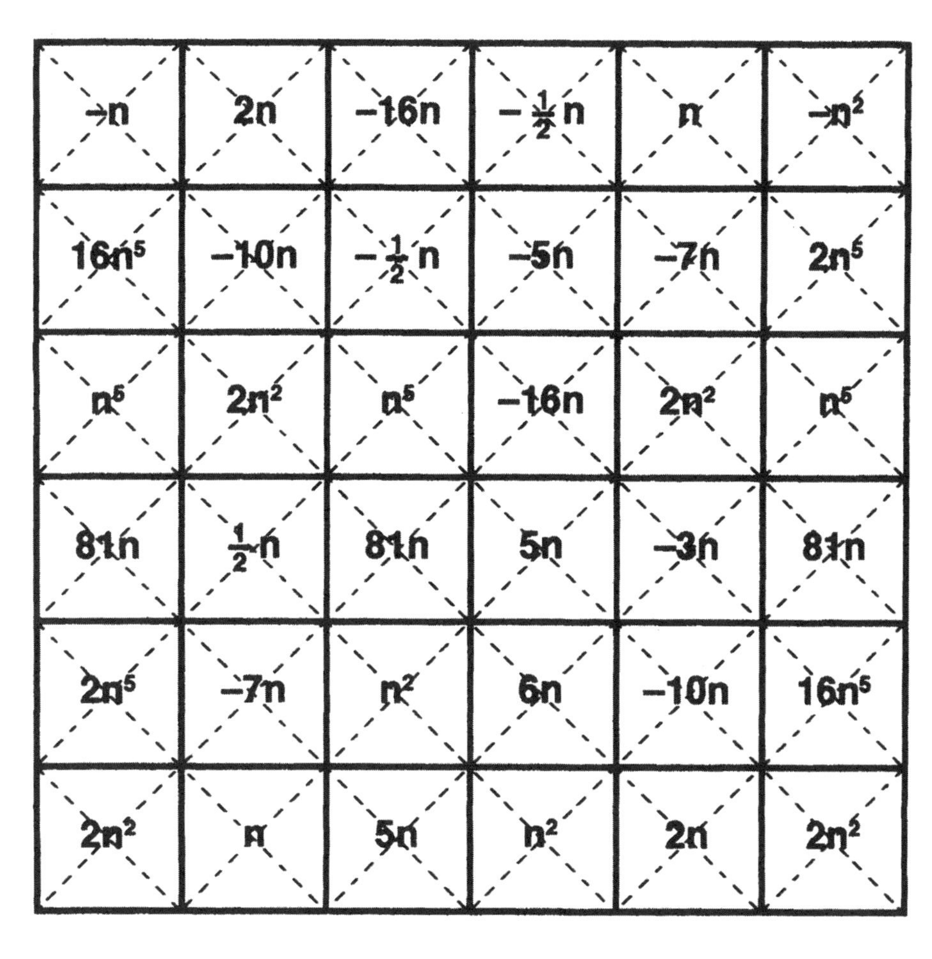

Find the next term in each sequence.

 $n, 2n, 3n, 4n,$?

$-n, -2n, -4n, -8n,$?

 $-2n, 4n, -6n, 8n,$?

 $n, -n, n, -n,$?

 $n, 3n, 9n, 27n,$?

 $n, -n^2, n^3, -n^4,$?

 $6n, 5n, 4n, 3n,$?

 $-8n, -4n, -2n, -n,$?

 $n, -2n^2, 4n^3, -8n^4,$?

 $-10n^2, -7n^2, -4n^2, -n^2,$?

 $13n, 8n, 3n, -2n,$?

 $-16n^2, -8n^2, -4n^2, -2n^2,$?

 $-6n, -3n, 0, 3n,$?

 $2n, -2n^2, 2n^3, -2n^4,$?

 $15n, 10n, 5n, 0,$?

ACTIVITY 31

Name ______________________

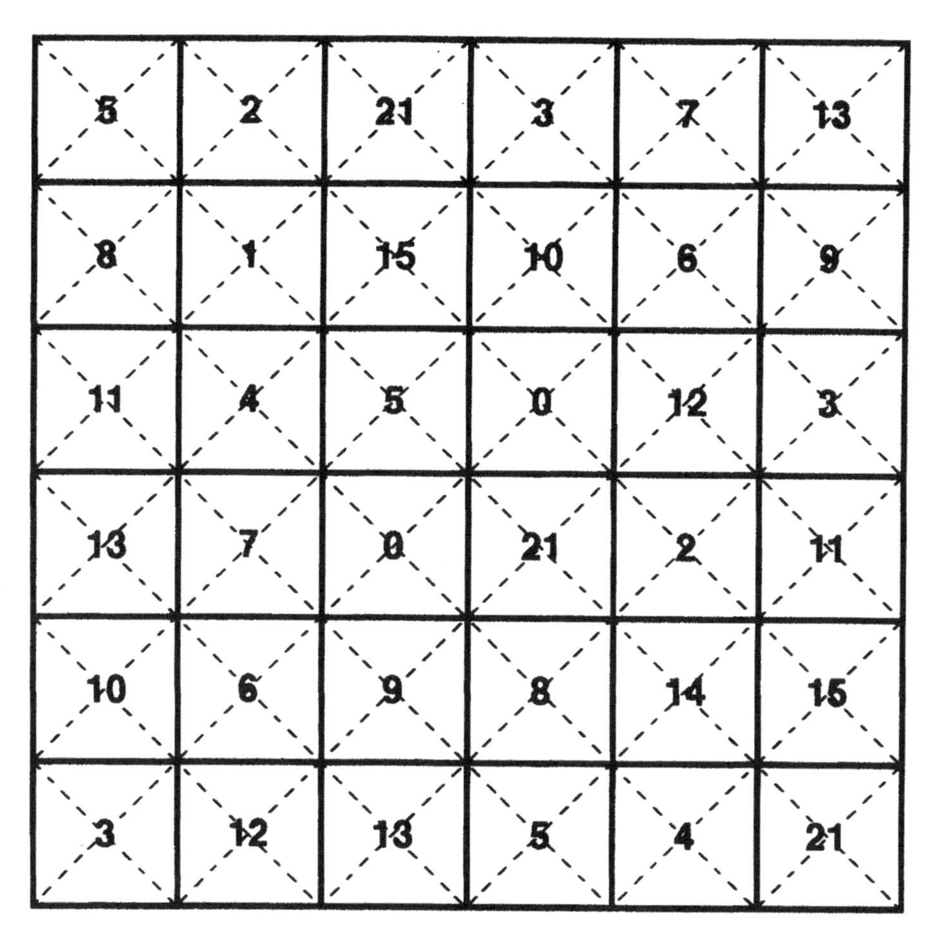

Solve for x in each equation.

$x - 3 = 1$

$5 + x = 12$

$9 = x + 3$

$x + 4 = 9$

$2x = 16$

$5 = x - 4$

$3 + x - 5 = 1$

$x - 3 = 2 - 3$

$\frac{x}{5} = 2$

$x - 8 = 7$

$11 + x = 24$

$18 = x - 3$

$2x = 22$

$5 + x - 12 = 5$

$-8 = 4 + x - 12$

Name ______________________

−1	−10	1	5	1	−8
−5	−8	−1	3	−1	−3
−9	−7	0	0	−9	−4
9	−4	2	2	−7	9
−5	−2	1	−10	7	−3
1	4	3	−1	4	5

Solve for x.

$x + 7 = 4$

$-4 = x + 3$

$-6 + x = -4$

$x - 5 = -1$

$8 + x - 12 = 1$

$-3 = 5 + x - 9$

$x + 6 = -3$

$4 - 7 + x = -4$

$-5 = -8 + x + 3$

$-11 + x = -8$

$-5 + x = 4$

$-2 = 3 + x$

$4 + x + 3 = -3$

$-12 + 7 = x + 3$

$-9 = -7 + x + 2$

ACTIVITY 33 Name ____________

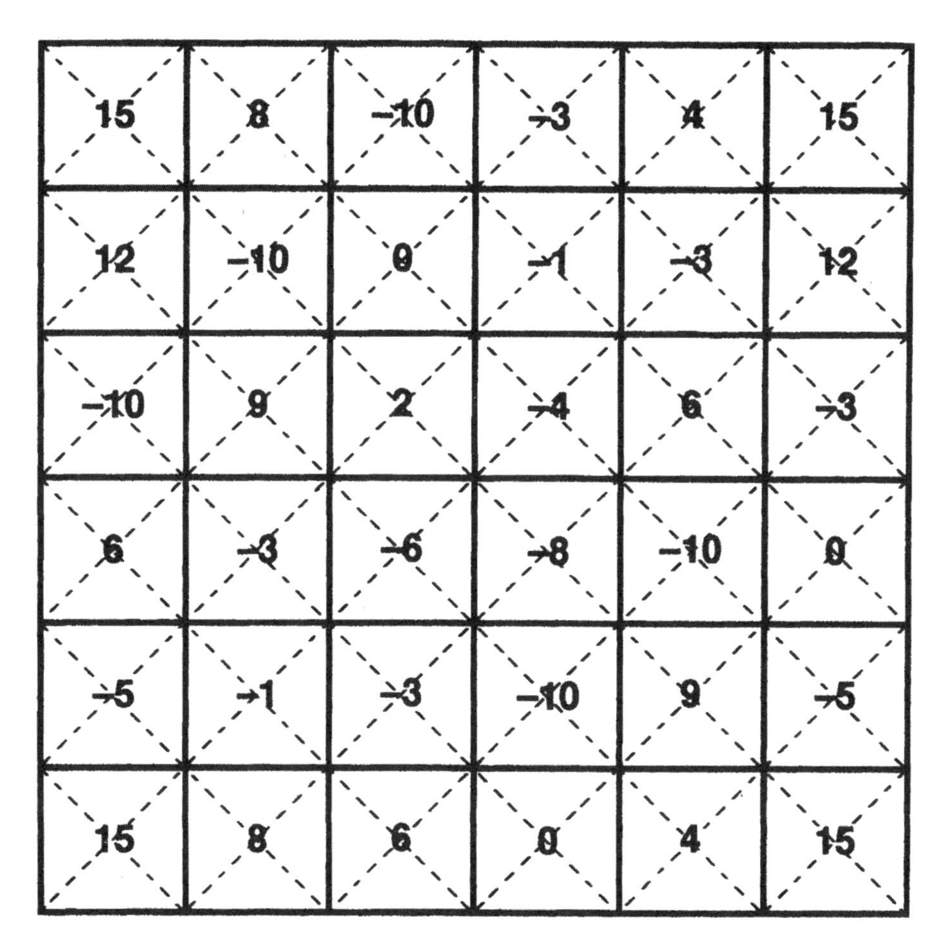

Solve for x.

$-3x = 15$

$\frac{x}{4} = -2$

$-30 = -5x$

$\frac{x}{-3} = -5$

$2x + 1 = 9$

$-3x - 1 = 11$

$10 = 5 - 5x$

$\frac{x}{2} + 1 = 5$

$6 = \frac{x}{3} + 3$

$10x - 2 = 18$

$\frac{2x}{3} = -2$

$5 = \frac{x}{4} + 2$

$-3 + \frac{x}{6} = -4$

$-8 = \frac{x}{7} - 8$

$-3 = \frac{2x}{5} + 1$

ACTIVITY 34 Name ____________________

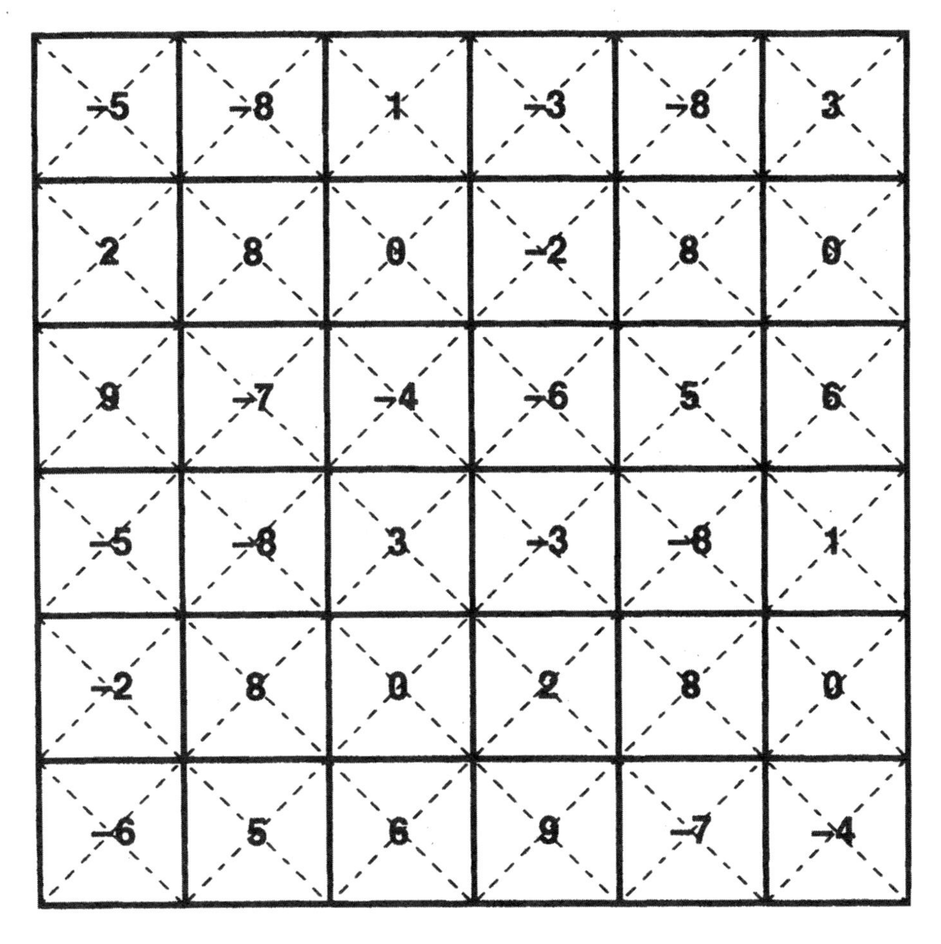

Solve for x.

$x + 2x = -12$

$3x - x = 10$

$-6 = x + x$

$-6 = 5x - 2x$

$x - 5x = -24$

$2x + 3 = 3x$

$4x - 5x = 5$

$5x - 5x = 2x$

$8 - 5x = -18$

$8x - 9x = -1$

$7x = -14 + 5x$

$-5 - 2x = 3 - 6x$

$-12x + 5x = 56$

$-6 + 7x - 6x = 3$

$4x - 2 = 3x + 6$

ACTIVITY 35

Name ______________________

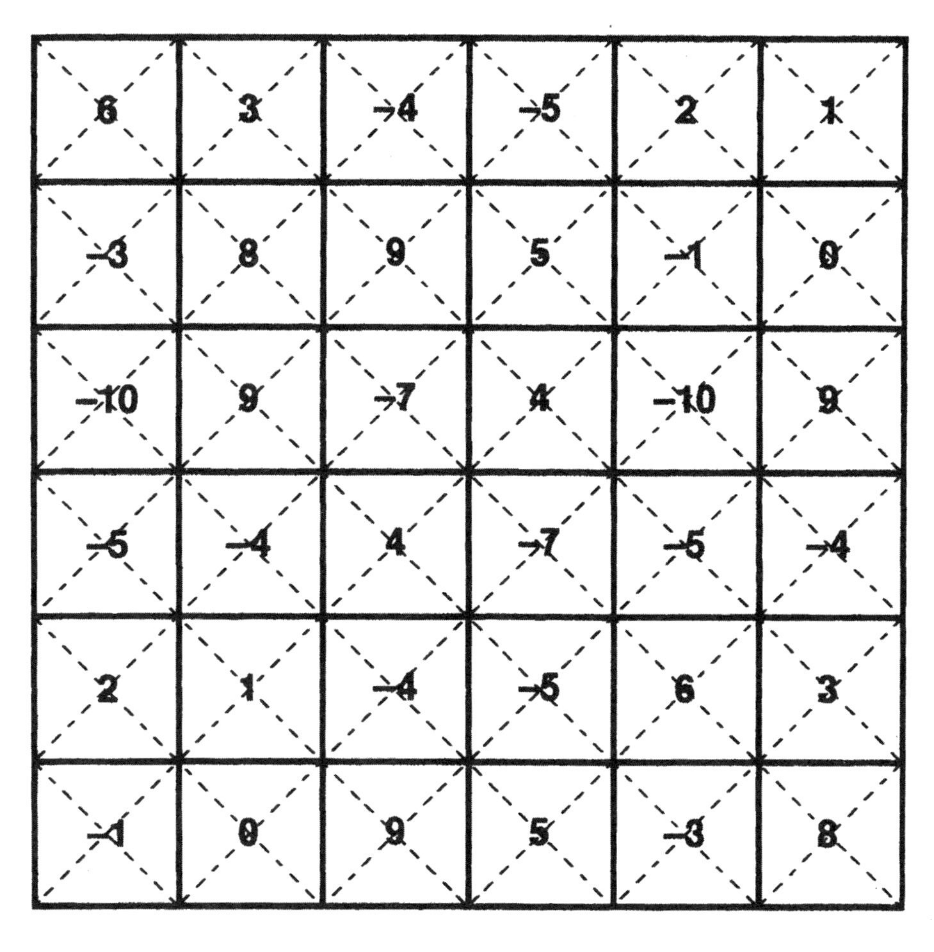

Solve for x.

Note: $-(-x) = x$.

 $5x - 6x = -4$

$4x - 3 = 5x + 7$

$-3 = -2x + x$

$-4 + 6x = -2x + 12$

 $3x - 4x = -6$

$3.5x - 4.5x = 4$

$x - 5x = 12$

$-x + 4 = 2x - 11$

$3x - 5x = 3 + x$

$4x - 5 = 5x - 5$

 $-3 - x = 11 + x$

 $-8x + 7x = -4 - 5$

 $12.5x - 8.5x = -20$

 $0 - 8 = 8x - 9x$

 $x - 4x + 2x = 6 - 7$

ACTIVITY 36

Name ______________________

Solve for x.

3.2	−0.2	5.2	6.5	2.2	1.6
3	1.1	−3	−1.1	2	−3.2
1.6	−2	−3.6	5.2	−2.1	−0.5
6.5	2.2	1.6	3.2	−0.2	5.2
−1.1	−1.6	−3.2	3	1.1	−3
5.2	−2.1	−0.5	1.6	−2	−3.6

$-2x + 3.2 = -2.8$

$3x - 1.2 = 5.4$

$4.7 - x = 8.3$

$-2.1x + 5.9 = 10.1$

$8.9 = 5x + 11.4$

$0.8 - 2x = -1.4$

$0.5x - 4.3 = -1.7$

$9 = -4x - 3.8$

$10.3 - 2x = -2.7$

$9.6 = 7.5 - x$

$4x - 0.8 = -12.8$

$3 - 5x = -5$

$-0.8 = x - 0.6$

$7.3 - 2x = 0.9$

$10x + 8.3 = -2.7$

Name ____________________

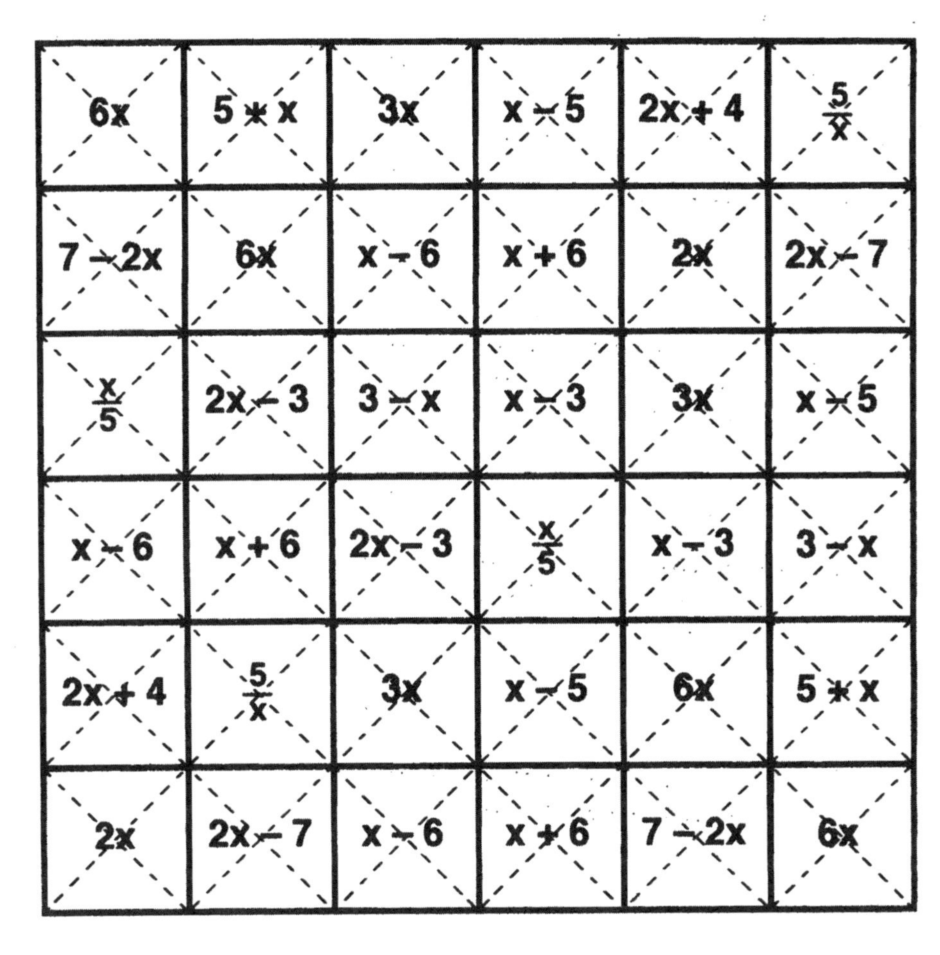

Write an algebraic expression for each word expression. Use x for the unknown number.

 Six more than an unknown number

 Twice an unknown number

 An unknown number decreased by six

 Five increased by an unknown number

 An unknown number tripled

 Four more than twice an unknown number

 Three less than an unknown number

 Three less an unknown number

 Twice an unknown number decreased by 3

 An unknown number diminished by five

 The product of six and an unknown number

 An unknown number divided by 5

 Five divided by an unknown number

 Seven decreased by twice an unknown number

 Seven less than twice an unknown number

ACTIVITY 38

Name ____________________

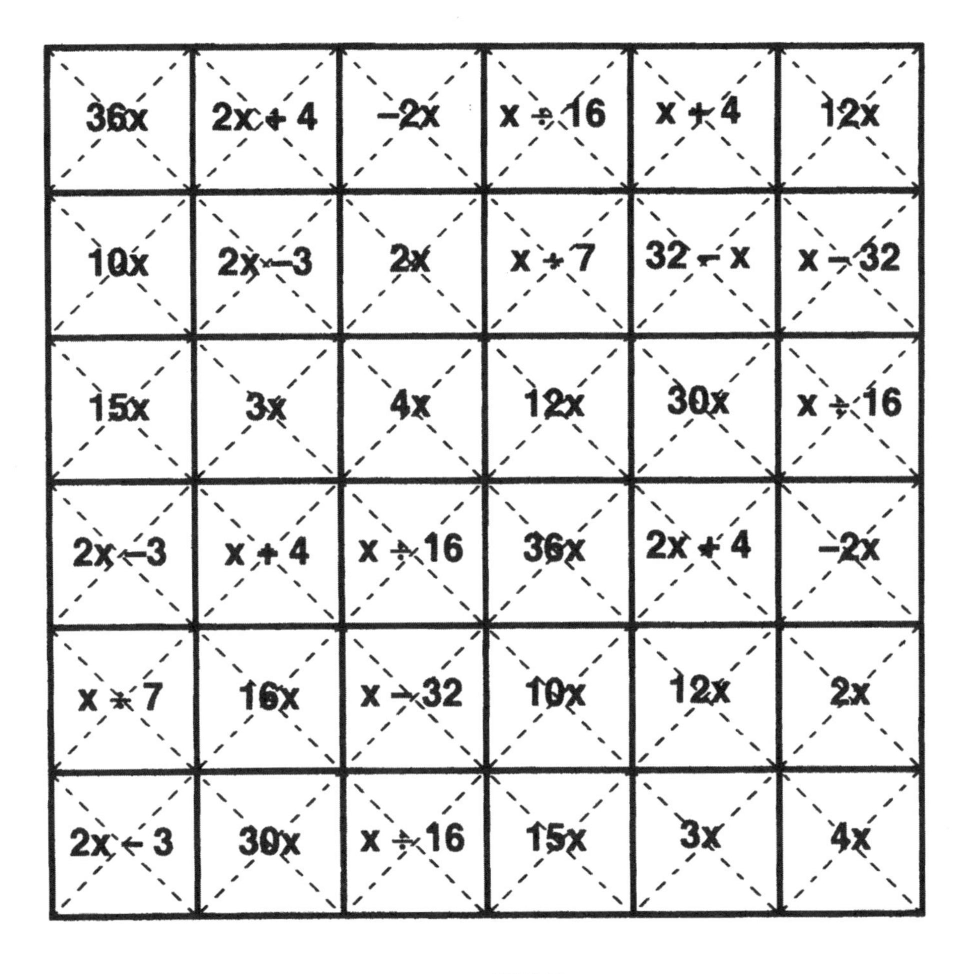

36x	2x + 4	−2x	x ÷ 16	x + 4	12x
10x	2x − 3	2x	x ÷ 7	32 − x	x − 32
15x	3x	4x	12x	30x	x ÷ 16
2x − 3	x + 4	x ÷ 16	36x	2x + 4	−2x
x ÷ 7	16x	x − 32	10x	12x	2x
2x − 3	30x	x ÷ 16	15x	3x	4x

Write an algebraic expression in simplest form for each word expression. Use x for each unknown number.

 **The sum of an unknown number and three less than the unknown number**

 **The sum of an unknown number and three times the unknown number**

 **The value in cents of an unknown number of dimes**

 **The number of inches in an unknown number of yards**

 **An unknown number subtracted from four times the unknown number**

 **The number of weeks equivalent to an unknown number of days**

 **Three times an unknown number decreased by five times the unknown number**

 **The number of pounds equivalent to an unknown number of ounces**

 **32 degrees subtracted from an unknown temperature**

 **The difference between a father's age and his son's age if the father's age is three times his son's age**

 **A girl's earnings for working 8 hours one day and 7 hours the next day at $x per hour**

 **The sum of the weights of twin brothers if one weighs four pounds more than the other**

 **The number of months equivalent to an unknown number of years**

 **The value in cents of a number of quarters and the same number of nickels**

 **The age of a older sister who is four years older than her younger sister**

Name ______________________

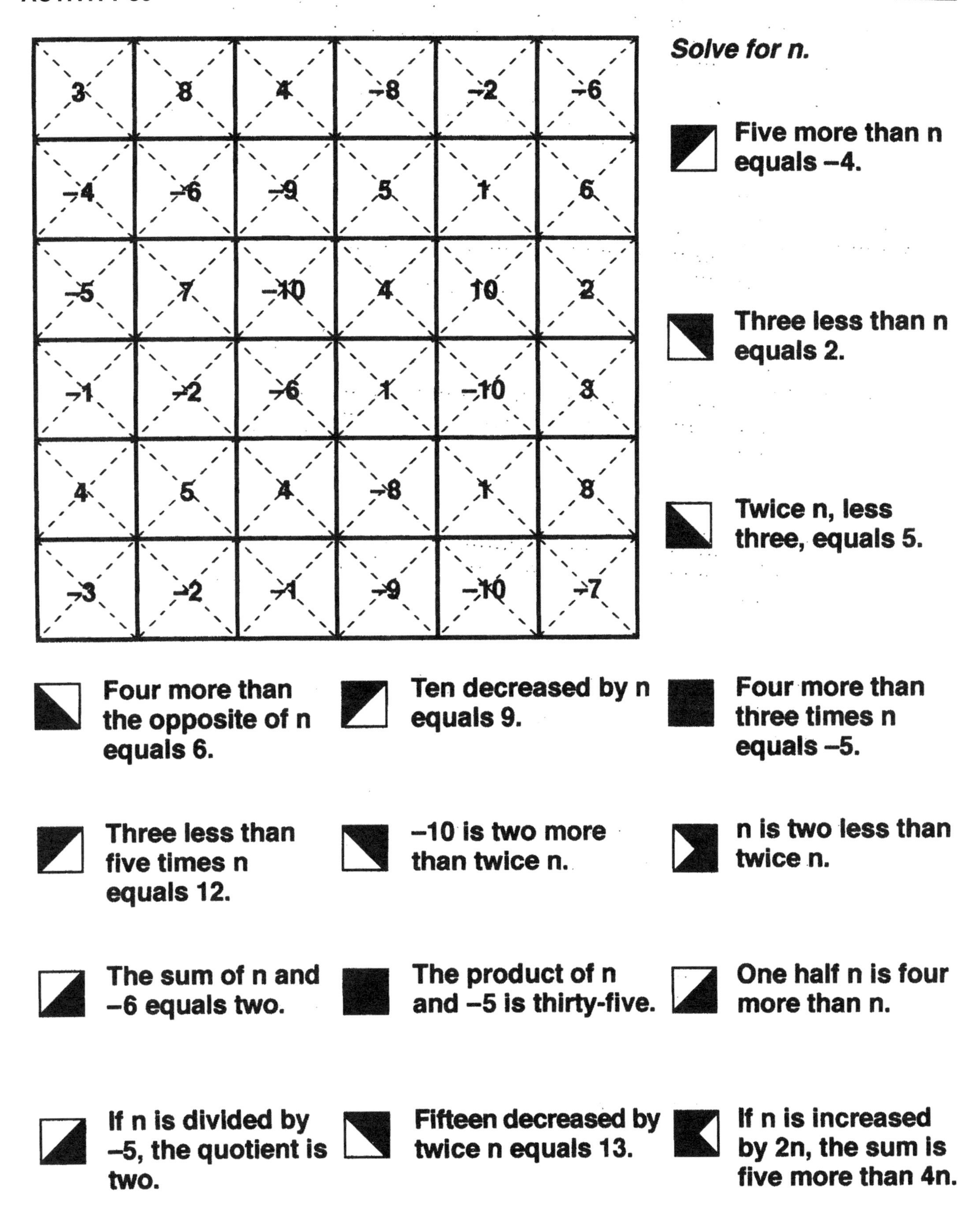

Solve for n.

Five more than n equals –4.

Three less than n equals 2.

Twice n, less three, equals 5.

Four more than the opposite of n equals 6.

Ten decreased by n equals 9.

Four more than three times n equals –5.

Three less than five times n equals 12.

–10 is two more than twice n.

n is two less than twice n.

The sum of n and –6 equals two.

The product of n and –5 is thirty-five.

One half n is four more than n.

If n is divided by –5, the quotient is two.

Fifteen decreased by twice n equals 13.

If n is increased by 2n, the sum is five more than 4n.

Name ______________________

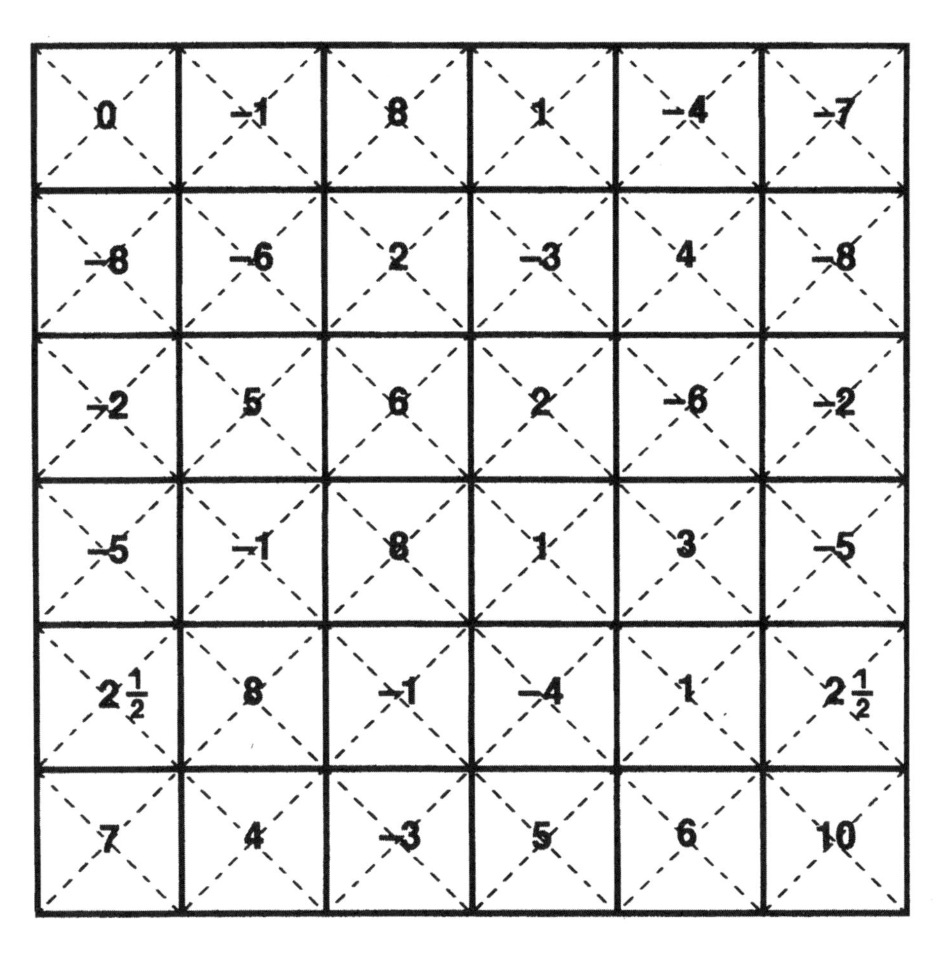

Solve for n.

 Two more than a number n equals –6.

 Three less than a number n equals 2.

 Twice a number n equals 12.

 Five increased by twice a number n equals –3.

 Five decreased by twice a number n equals –3.

 Three less a number n equals 2.

 Five is three more than a number n.

 Seven is three less than four times a number n.

 Twice a number n diminished by three equals –9.

 Three is five more than a number n.

 Two less than a number n equals –3.

 Six equals the triple of a number n decreased by 3.

 One-half a number n, less five, equals –1.

 Two is five more than one-half a number n.

 Three times a number n increased by ten equals –5.

ACTIVITY 41 Name ____________________

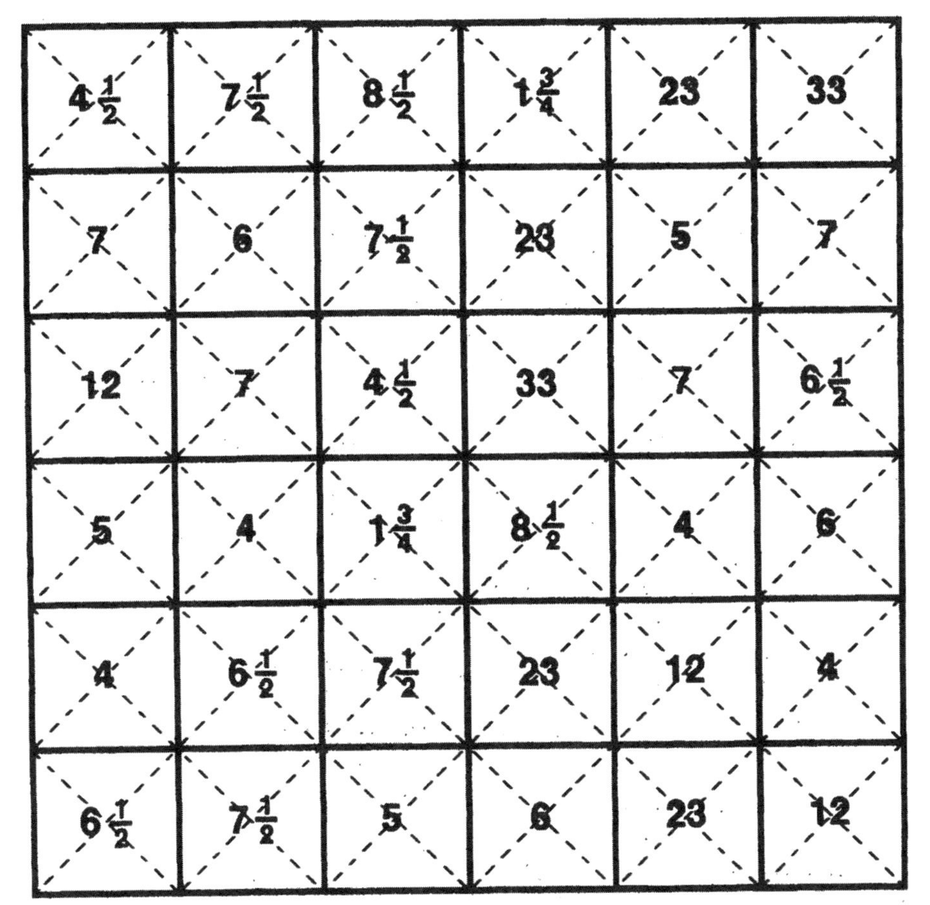

One team scores three more than twice the number of runs as the other team. The winners score 11 runs. What is the loser's score?

A pair of socks and a shirt cost $40. The cost of the shirt was $5 more than four times the cost of the socks. What did the socks cost?

On Saturday Harold worked three less hours than on Friday. If he worked 10 hours in the two days, how many did he work on Friday?

Jan scored six more points in a game than Bill. Together they scored 18 points. How many points did Jan score?

On a trip that took 10 hours, Mark drove 2 fewer hours than Mary. How many hours did Mary drive?

Ruth is twice the age of her brother. The sum of their ages is 15. How old is Ruth's brother?

Don and Ron together weighed 48 pounds. Don weighed two pounds more than Ron. How much did Ron weigh?

A rectangle's length is 4 inches more than its width. The sum of the length and width is 13 inches. What is the width?

Mike worked 2 more hours on Saturday than on Friday. He worked 15 hours in the two days. How long did he work on Saturday?

Ann and Tim saw two movies in one day. The two movies lasted 4 hours. One movie was half an hour longer than the other. How many hours did the shorter movie run?

The high temperature one day was twice the low temperature. The high temperature was also 33 degrees more than the low temperature. What was the low temperature?

Some hikers made the round trip on a mountain trail in 10 hours. It took three times longer to hike up than to hike down. How many hours did it take to hike up the trail?

ACTIVITY 42 Name ____________

$10x + 5$	$-4x + 8$	$6x - 3$	$4x + 4$	$-2x + 5$	$2x + 6$
$2x + 10$	$10x + 6$	$3x + 12$	$4x - 4$	$6x - 10$	$-3 + x$
$2x + 6$	$3x - 3$	$4x + 4$	$10x - 5$	$-3x + 3$	$10x + 5$
$4x + 4$	$-2x + 5$	$2x + 6$	$10x + 5$	$-4x + 8$	$6x - 3$
$4x - 4$	$6x - 10$	$-3 + x$	$2x + 10$	$10x + 6$	$3x + 12$
$10x - 5$	$-3x + 3$	$10x + 5$	$2x + 6$	$3x - 3$	$4x + 4$

Multiply.

$2(x + 5)$

$-3(x - 1)$

$2(5x + 3)$

$-5(-2x + 1)$

$(x - 1)4$

$-5(-2x - 1)$

$6(\frac{1}{2}x + 2)$

$-\frac{1}{3}(9 - 3x)$

$(3x - 5)2$

$2(2x + 2)$

$(-2x + 1)(-3)$

$-4(x - 2)$

$-1(-2x - 6)$

$3(x - 1)$

$\frac{1}{2}(-4x + 10)$

ACTIVITY 43

Name ______________________

$2n^2-n$	n^3+n^2	$-3n-3n^2$	$-3n-3n^2$	n^3-n	$n+n^4$
n^2-n	$2n^2-n$	$2n^5-2n$	$-2n+n^2$	$n+n^4$	$-6n+n^2$
$3n^2-3n$	$2n^2+2n$	n^2+4n	n^2+n	$-n^2+3n$	$3n^3-n^2$
$3n^3-n^2$	$-6n+2n^2$	$n+n^4$	$2n^2-n$	n^2-n	$3n^2-3n$
$2n^2+2n$	n^2+n	n^3-n	n^3+n^2	n^2+4n	$-n^2+3n$
n^2+n	$-2n+n^2$	$-3n-3n^2$	$-3n-3n^2$	$2n^5-2n$	n^2+4n

Multiply.

$2n(n+1)$

$3n(n-1)$

$(n+4)n$

$-n(n-3)$

$\frac{1}{2}n(2n-2)$

$n(n^2+n)$

$-3n(1+n)$

$(6n-3)\frac{1}{3}n$

$n(n^2-1)$

$-\frac{1}{4}n(8-4n)$

$(3n^2-n)n$

$-n(-n-1)$

$n(1+n^3)$

$-2n(3-n)$

$(n^4-1)2n$

ACTIVITY 44

Name ____________________

$-\frac{3}{8}$	$\frac{1}{5}$	$\frac{2}{7}$	$\frac{2}{7}$	$-\frac{2}{9}$	$\frac{2}{3}$
$\frac{7}{9}$	$\frac{1}{3}$	$\frac{4}{9}$	$-\frac{4}{9}$	$-\frac{2}{3}$	$-\frac{1}{8}$
$-\frac{10}{11}$	$\frac{2}{3}$	$-\frac{2}{9}$	$\frac{5}{12}$	$-\frac{5}{12}$	$\frac{1}{5}$
$-\frac{3}{8}$	$\frac{5}{12}$	$\frac{4}{9}$	$-\frac{4}{9}$	$-\frac{10}{11}$	$\frac{2}{3}$
$\frac{7}{9}$	$\frac{1}{3}$	$-\frac{2}{9}$	$\frac{1}{5}$	$-\frac{2}{3}$	$-\frac{1}{8}$
$-\frac{10}{11}$	$-\frac{4}{9}$	$-\frac{2}{7}$	$-\frac{2}{7}$	$-\frac{5}{12}$	$\frac{5}{12}$

Add or Subtract.

Remember: $\frac{-1}{2} = -\frac{1}{2}$

$\frac{-a}{b} = -\frac{a}{b}$

$\frac{2}{9} + \frac{5}{9}$

$\frac{5}{3} - \frac{4}{3}$

$\frac{3}{8} - \frac{4}{8}$

$\frac{-1}{12} + \frac{-4}{12}$

$\frac{8}{9} - \frac{10}{9}$

$\frac{-3}{5} + \frac{4}{5}$

$-\frac{2}{3} + \frac{4}{3}$

$\left(-\frac{4}{11}\right) + \left(-\frac{6}{11}\right)$

$\frac{6}{7} - \frac{4}{7}$

$\left(-\frac{1}{9}\right) - \left(\frac{3}{9}\right)$

$\left(-\frac{7}{8}\right) + \frac{4}{8}$

$\left(-\frac{2}{9}\right) + \frac{6}{9}$

$\left(\frac{1}{12}\right) - \left(\frac{-4}{12}\right)$

$\left(-\frac{1}{7}\right) + \left(-\frac{1}{7}\right)$

$\left(-\frac{7}{3}\right) + \left(\frac{5}{3}\right)$

Name ____________________

$\frac{2}{3n}$	$\frac{-2}{n}$	$\frac{-n}{3}$	$\frac{-4n}{5}$	$\frac{2n}{5}$	$\frac{13n}{7}$
$\frac{3n}{4}$	$\frac{2}{5n}$	$\frac{1}{2n}$	$\frac{7}{2n}$	$\frac{2}{5n}$	$\frac{-3n}{7}$
$\frac{2}{n}$	$\frac{13}{7n}$	$\frac{5}{n}$	$\frac{-4n}{5}$	$\frac{-n}{3}$	$\frac{4n}{7}$
$\frac{3n}{4}$	$\frac{13n}{7}$	$\frac{7}{2n}$	$\frac{1}{2n}$	$\frac{2}{3n}$	$\frac{4n}{7}$
$\frac{2}{n}$	$\frac{2n}{5}$	$\frac{4n}{7}$	$\frac{2}{n}$	$\frac{-2}{n}$	$\frac{-3n}{7}$
$\frac{2n}{5}$	$\frac{13n}{7}$	$\frac{5}{n}$	$\frac{-2}{n}$	$\frac{7}{2n}$	$\frac{13}{7n}$

Add or subtract.

$\frac{n}{4} + \frac{2n}{4}$

$\frac{5n}{7} - \frac{n}{7}$

$\frac{3}{n} + \frac{2}{n}$

$\frac{5}{2n} - \frac{4}{2n}$

$\frac{2n}{5} - \frac{6n}{5}$

$\frac{5}{2n} + \frac{2}{2n}$

$\frac{n}{3} - \frac{2n}{3}$

$\frac{-2n}{7} + \frac{-n}{7}$

$\frac{n}{5} + \frac{n}{5}$

$\frac{3}{n} - \frac{5}{n}$

$\frac{7}{3n} - \frac{5}{3n}$

$\frac{-3}{n} + \frac{5}{n}$

$\frac{-4}{5n} - \frac{-6}{5n}$

$\frac{11}{7n} - \frac{-2}{7n}$

$\frac{15n}{7} - \frac{2n}{7}$

ACTIVITY 46

Name ____________________

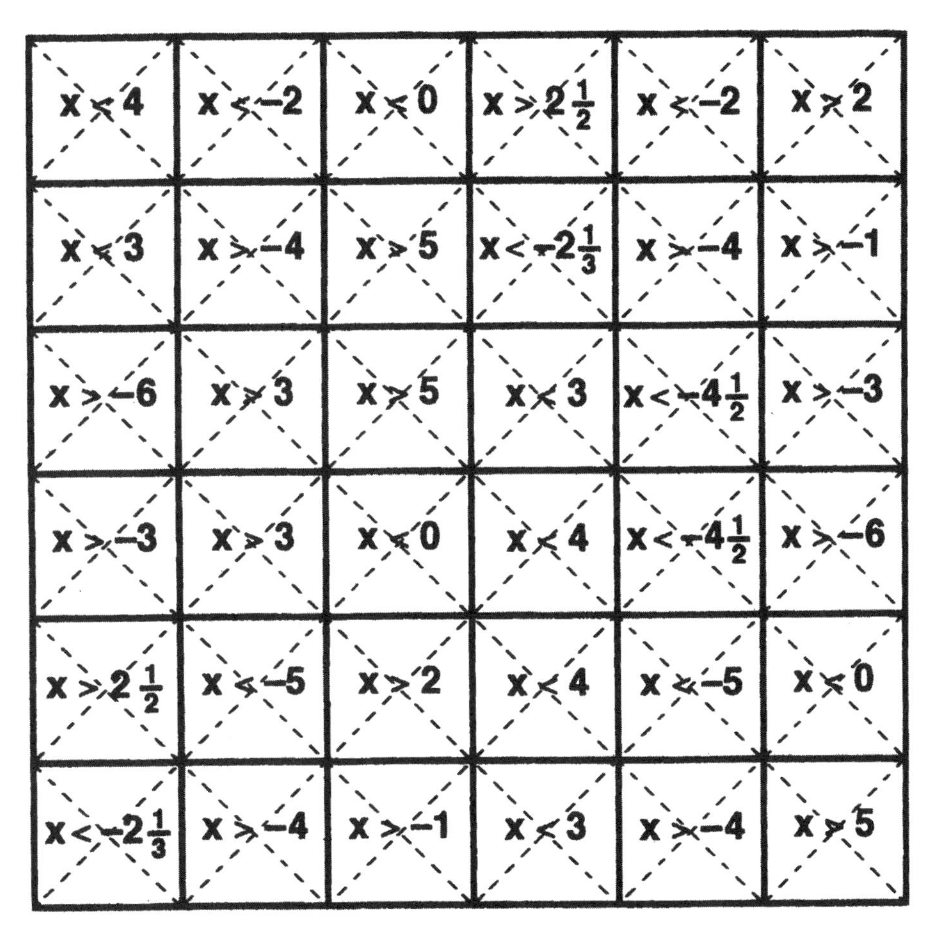

$x < 4$	$x < -2$	$x < 0$	$x > 2\frac{1}{2}$	$x < -2$	$x > 2$
$x < 3$	$x > -4$	$x > 5$	$x < -2\frac{1}{3}$	$x > -4$	$x > -1$
$x > -6$	$x > 3$	$x > 5$	$x < 3$	$x < -4\frac{1}{2}$	$x > -3$
$x > -3$	$x > 3$	$x < 0$	$x < 4$	$x < -4\frac{1}{2}$	$x > -6$
$x > 2\frac{1}{2}$	$x < -5$	$x > 2$	$x < 4$	$x < -5$	$x < 0$
$x < -2\frac{1}{3}$	$x > -4$	$x > -1$	$x < 3$	$x > -4$	$x > 5$

Solve for x.

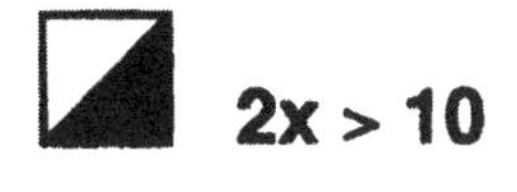

$2x > 10$

$x + 5 > 2$

$x - 6 < -3$

$3x < -6$

$2x + 1 > 5$

$-2 + x < 2$

$5x > -20$

$3x - 1 > -4$

$2x < -9$

$4x - 5 > -29$

$2 + 3x < -5$

$5 + 6x < -25$

$-3 + 9x > 24$

$4x - 2 > 8$

$2x - 6 < -6$

ACTIVITY 47

Name ______________________

$x < \frac{1}{3}$	$x < -6$	$x < 3$	$x < \frac{1}{3}$	$x < -4$	$x < 3$
$x < -3$	$x < -4$	$x < 2$	$x < -1$	$x < -6$	$x > -4$
$x < 1\frac{1}{2}$	$x < 0$	$x > 3$	$x > -2$	$x < \frac{1}{2}$	$x < 1\frac{1}{2}$
$x > 0$	$x < 1\frac{1}{2}$	$x < 1$	$x < -2$	$x < 1\frac{1}{2}$	$x > 0$
$x > -3$	$x > -1$	$x > -2$	$x > 4$	$x > \frac{1}{2}$	$x > 3$
$x < 4$	$x > -3$	$x < 1$	$x < -2$	$x > 4$	$x < 4$

Solve for x. Hint: $3 < x$ means the same as $x > 3$.

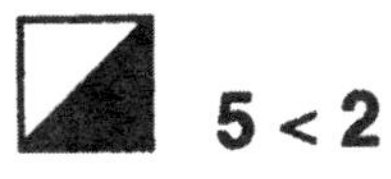

$5 < 2 + x$

$-4 > x - 2$

$5 > x + 4$

$-3 < 2x + 1$

$11 > 2 + 3x$

$-1 > 5 + x$

$-5 < x - 5$

$x + 4 < 2x$

$3 - x > x$

$2 - x < 3x$

$12 - 2x > x$

$-5 + x < 6x$

$x + 1 > 4x$

$2x - 9 < 5x$

$x - 12 > 4x$

 Name ______________________________

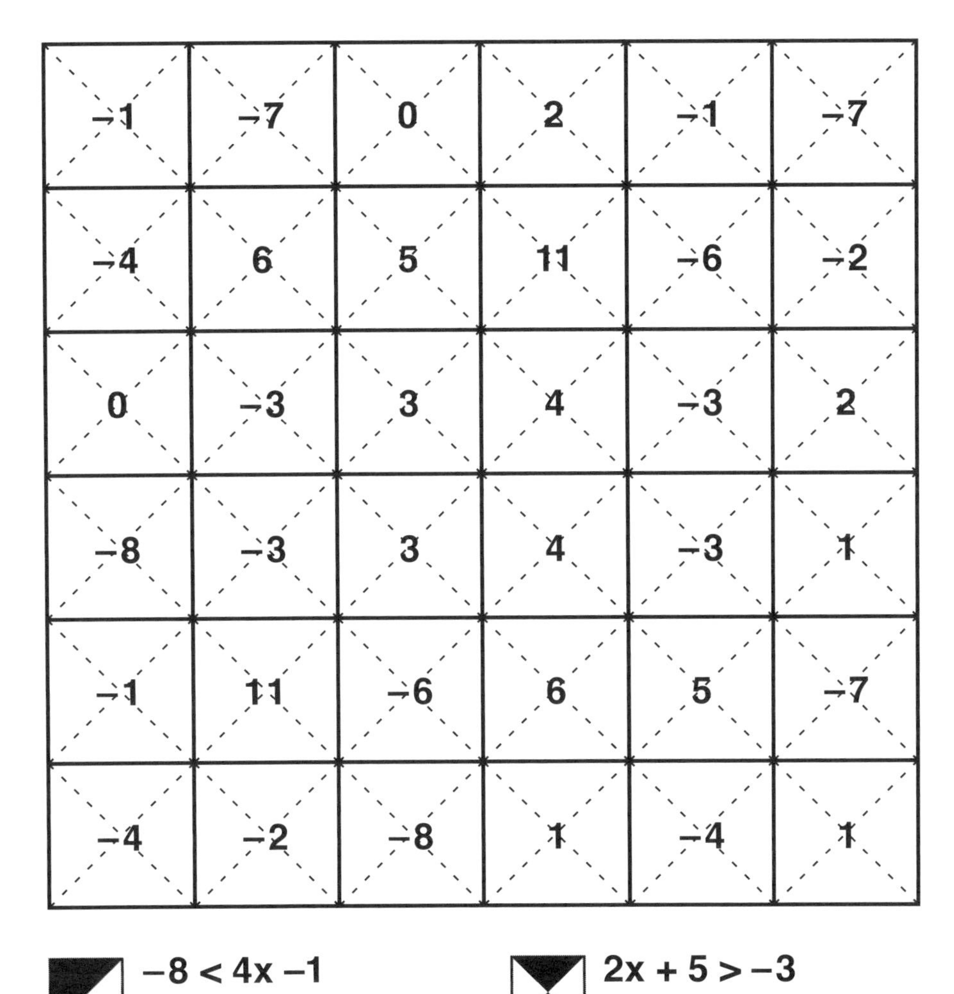

Find the least integral value of x to make each inequality true.

≤ means "less than *or* equal to." ≥ means "greater than *or* equal to."

$x + 2 > 7$

$-2 < x + 1$

$1 + 2x > 6$

$-8 < 4x - 1$

$2x + 5 > -3$

$x - 2 \geq -8$

$-3 < x - 3$

$12 \leq 2x + 4$

$3x \geq 3 + x$

$2x - 1 > x + 3$

$-3x - 1 < x$

$x \geq -x - 9$

$3x \geq x - 15$

$x + 21 < 3x$

$x - 17 \leq 3x$

ACTIVITY 1

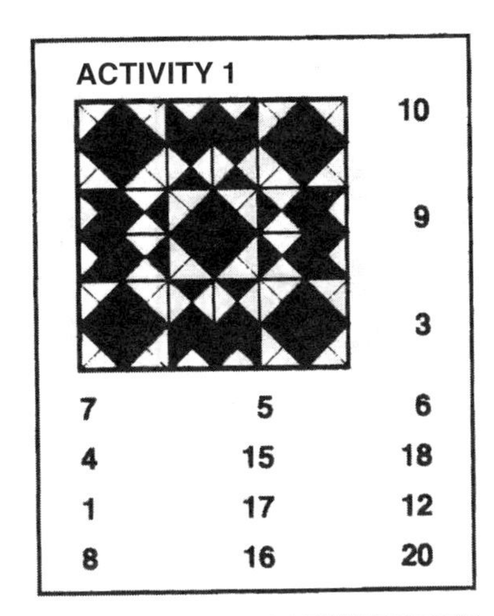

10

9

3

7	5	6
4	15	18
1	17	12
8	16	20

ACTIVITY 2

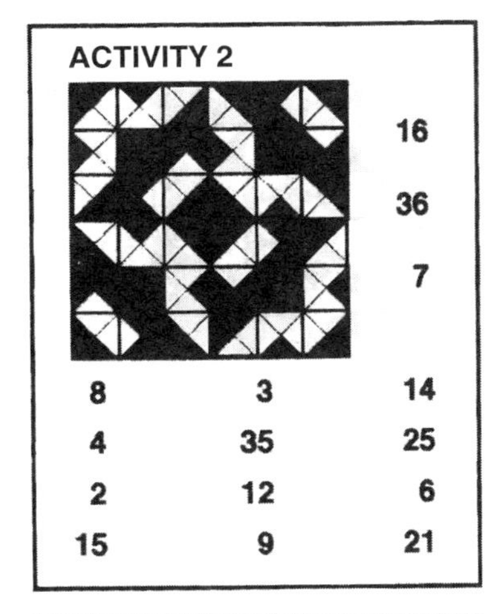

16

36

7

8	3	14
4	35	25
2	12	6
15	9	21

ACTIVITY 3

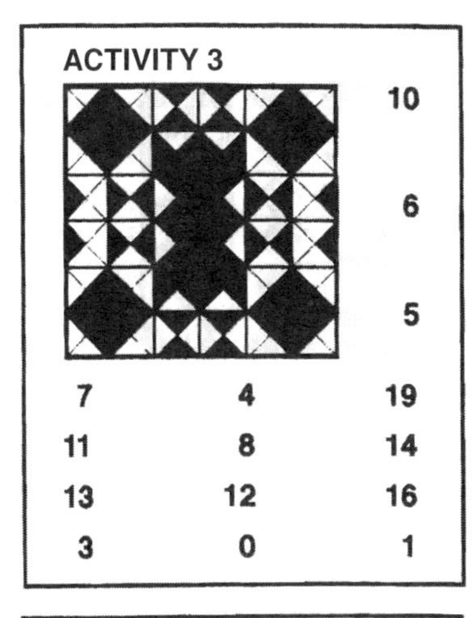

10

6

5

7	4	19
11	8	14
13	12	16
3	0	1

ACTIVITY 4

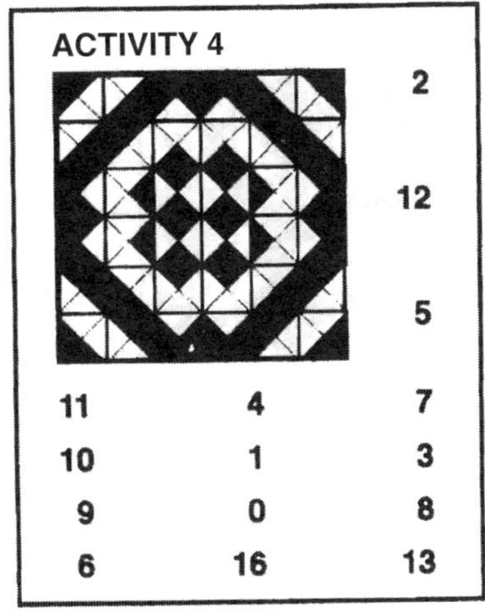

2

12

5

11	4	7
10	1	3
9	0	8
6	16	13

ACTIVITY 5

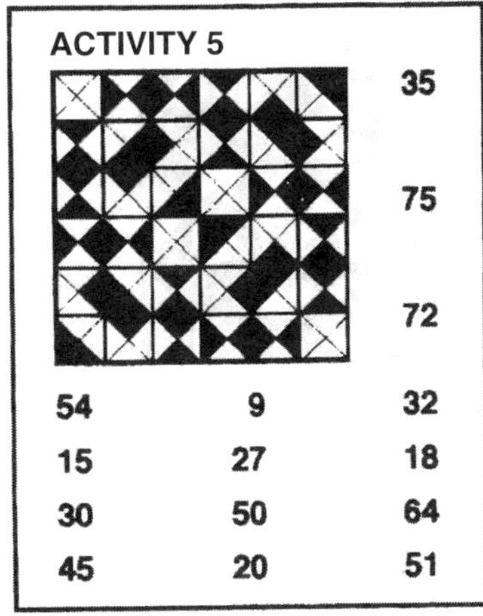

35

75

72

54	9	32
15	27	18
30	50	64
45	20	51

ACTIVITY 6

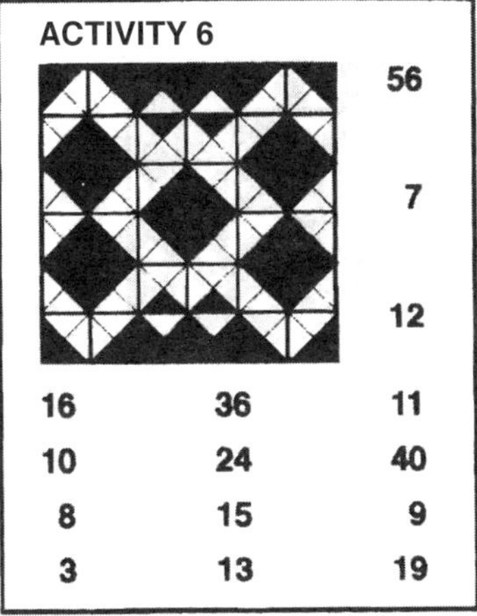

56

7

12

16	36	11
10	24	40
8	15	9
3	13	19

ACTIVITY 7

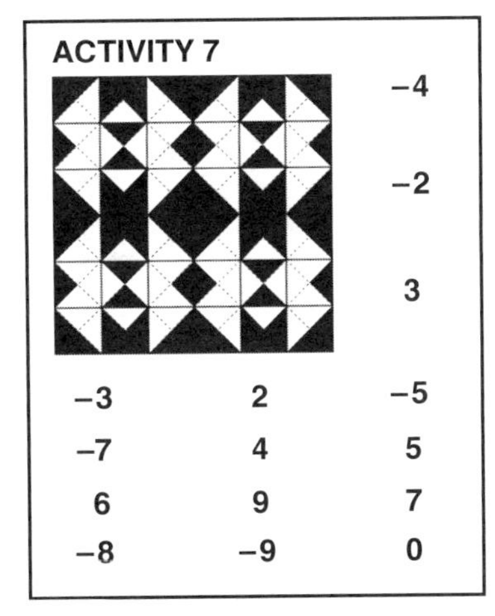

–4

–2

3

–3	2	–5
–7	4	5
6	9	7
–8	–9	0

ACTIVITY 8

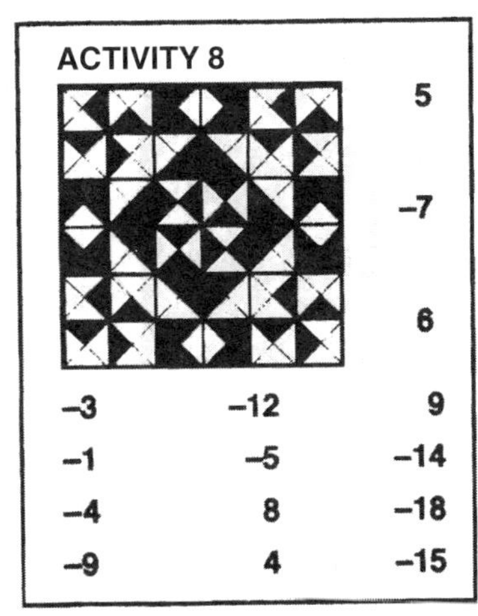

5

–7

6

–3	–12	9
–1	–5	–14
–4	8	–18
–9	4	–15

ACTIVITY 9

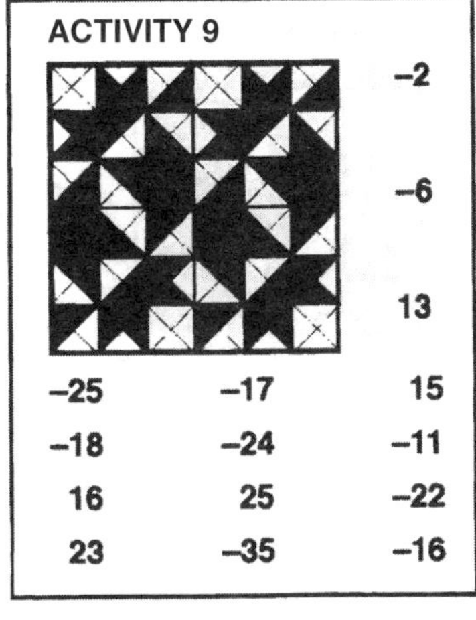

–2

–6

13

–25	–17	15
–18	–24	–11
16	25	–22
23	–35	–16

ACTIVITY 10

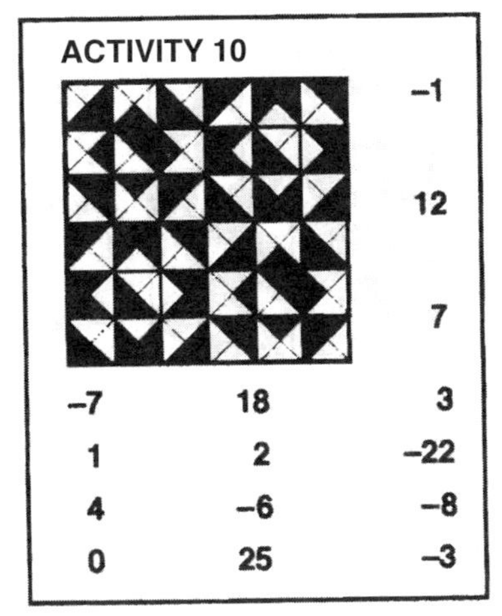

–1

12

7

–7	18	3
1	2	–22
4	–6	–8
0	25	–3

ACTIVITY 11

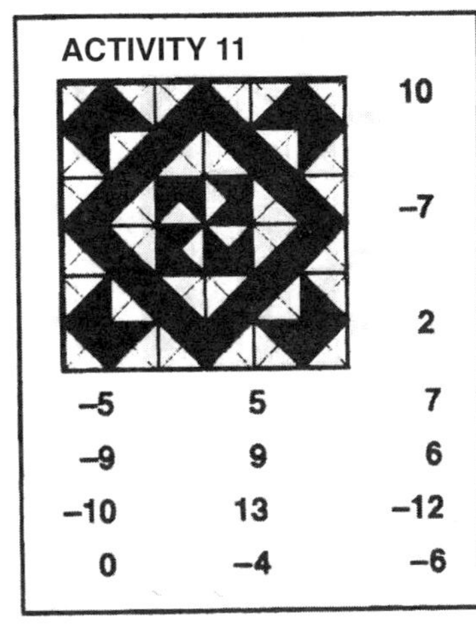

10

–7

2

–5	5	7
–9	9	6
–10	13	–12
0	–4	–6

ACTIVITY 12

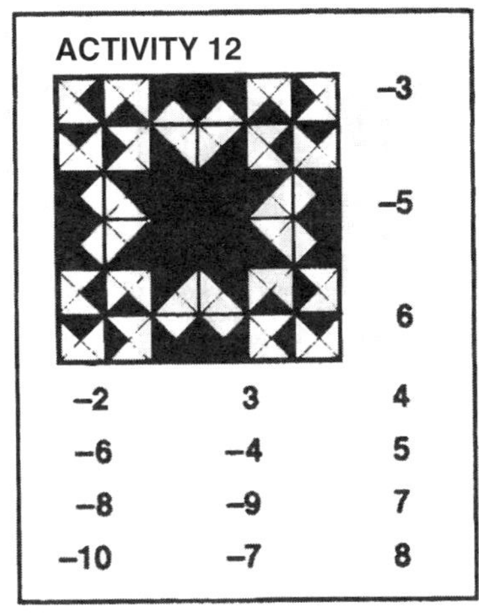

–3

–5

6

–2	3	4
–6	–4	5
–8	–9	7
–10	–7	8

ACTIVITY 13

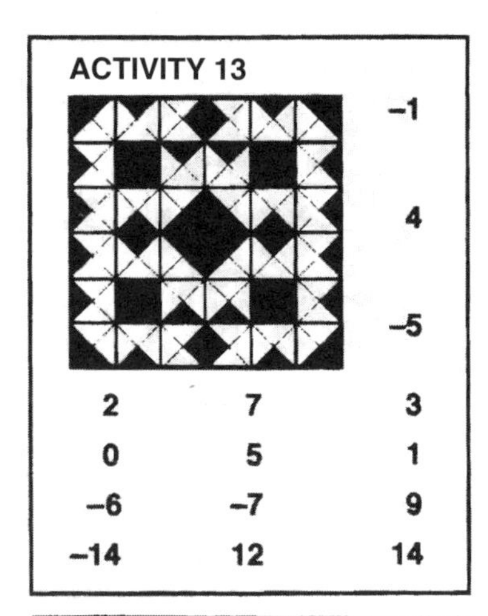

−1, 4, −5

2	7	3
0	5	1
−6	−7	9
−14	12	14

ACTIVITY 14

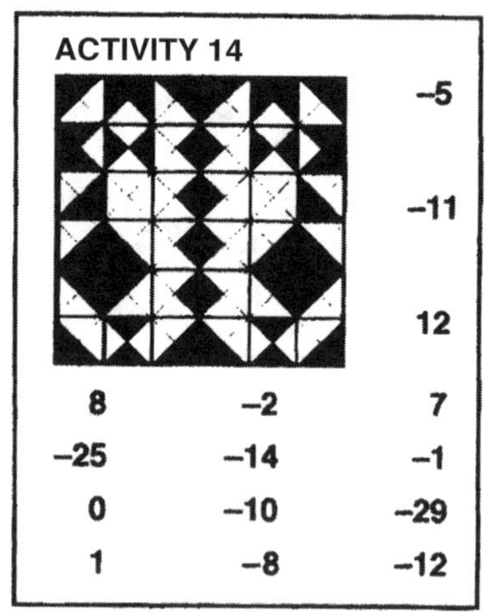

−5, −11, 12

8	−2	7
−25	−14	−1
0	−10	−29
1	−8	−12

ACTIVITY 15

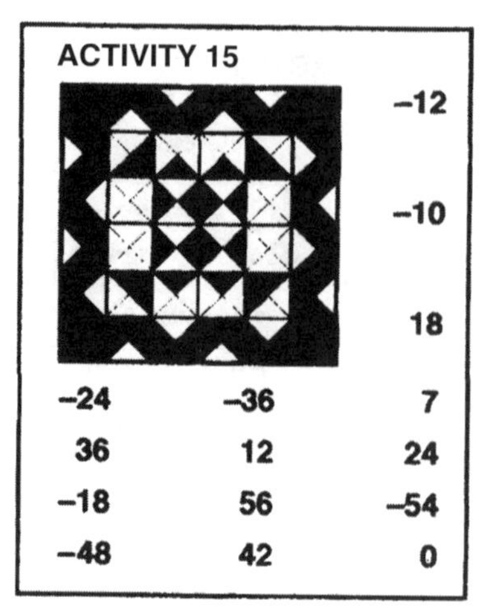

−12, −10, 18

−24	−36	7
36	12	24
−18	56	−54
−48	42	0

ACTIVITY 16

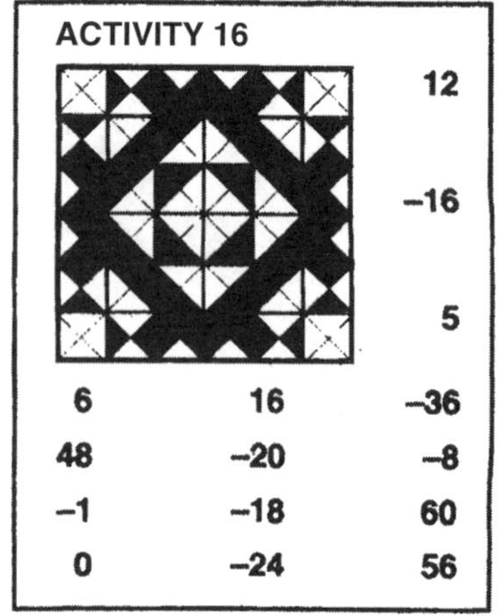

12, −16, 5

6	16	−36
48	−20	−8
−1	−18	60
0	−24	56

ACTIVITY 17

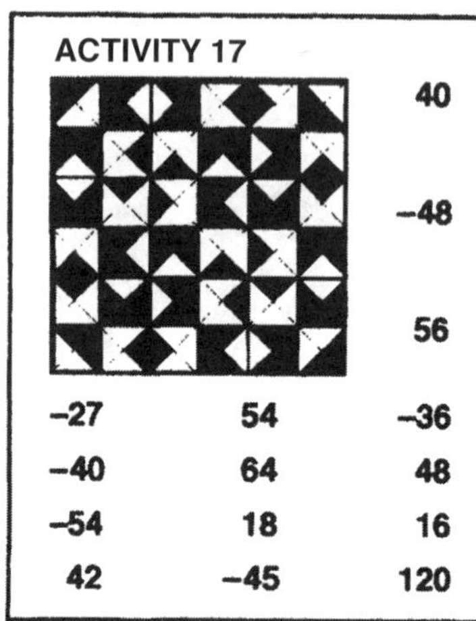

40, −48, 56

−27	54	−36
−40	64	48
−54	18	16
42	−45	120

ACTIVITY 18

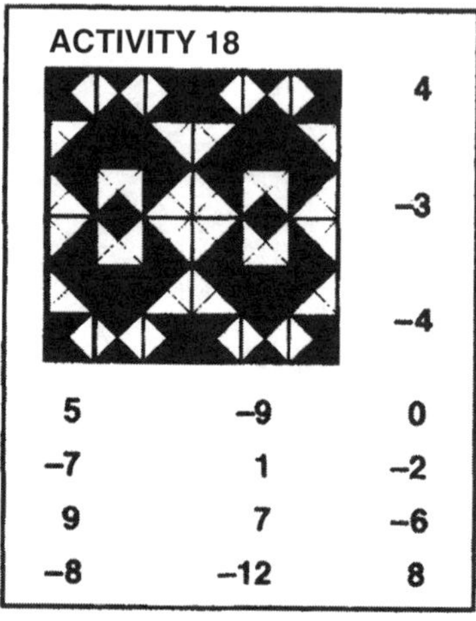

4, −3, −4

5	−9	0
−7	1	−2
9	7	−6
−8	−12	8

ACTIVITY 19

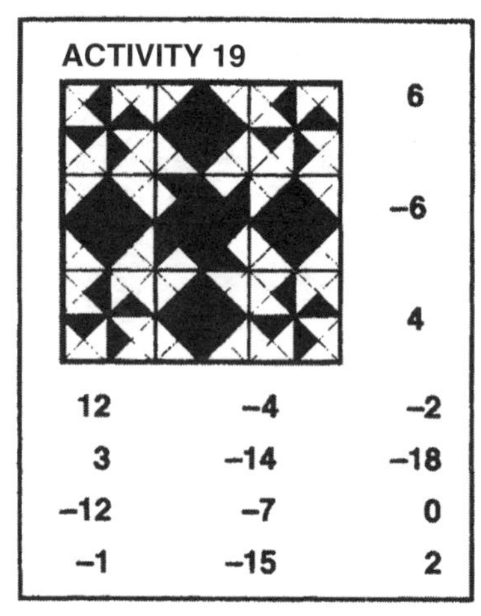

6, −6, 4

12	−4	−2
3	−14	−18
−12	−7	0
−1	−15	2

ACTIVITY 20

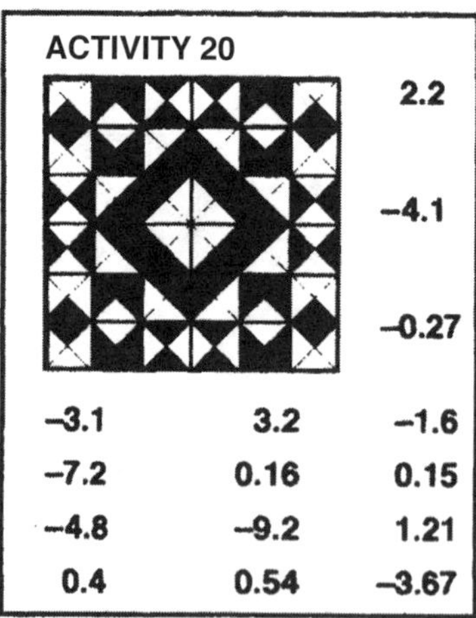

2.2, −4.1, −0.27

−3.1	3.2	−1.6
−7.2	0.16	0.15
−4.8	−9.2	1.21
0.4	0.54	−3.67

ACTIVITY 21

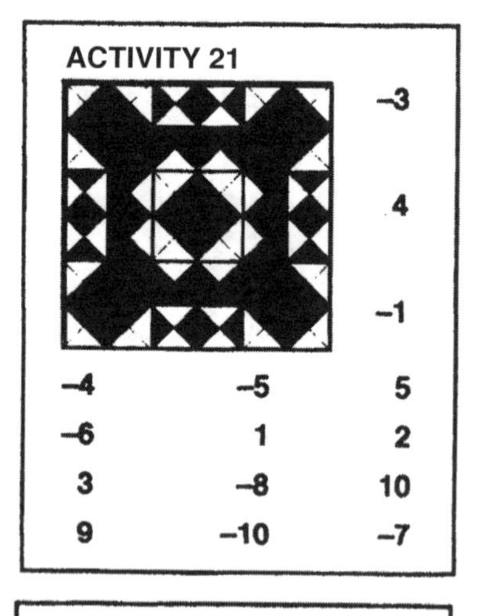

−3, 4, −1

−4	−5	5
−6	1	2
3	−8	10
9	−10	−7

ACTIVITY 22

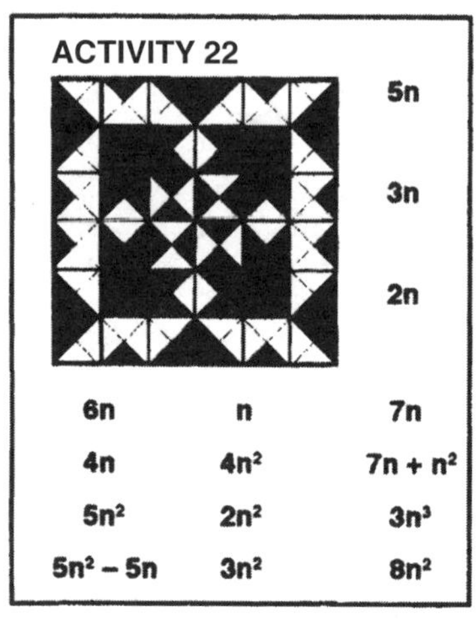

$5n$, $3n$, $2n$

$6n$	n	$7n$
$4n$	$4n^2$	$7n + n^2$
$5n^2$	$2n^2$	$3n^3$
$5n^2 - 5n$	$3n^2$	$8n^2$

ACTIVITY 23

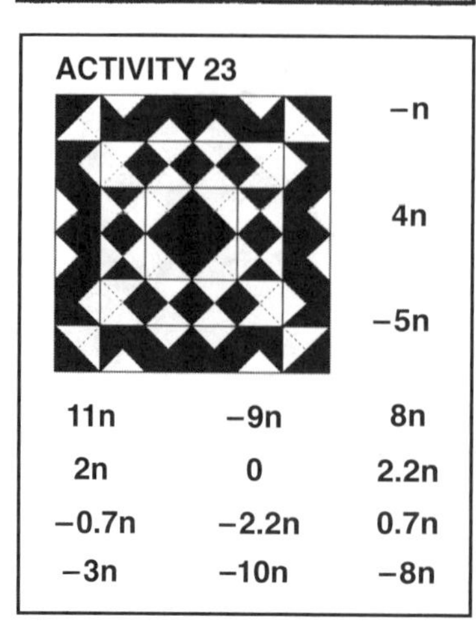

$-n$, $4n$, $-5n$

$11n$	$-9n$	$8n$
$2n$	0	$2.2n$
$-0.7n$	$-2.2n$	$0.7n$
$-3n$	$-10n$	$-8n$

ACTIVITY 24

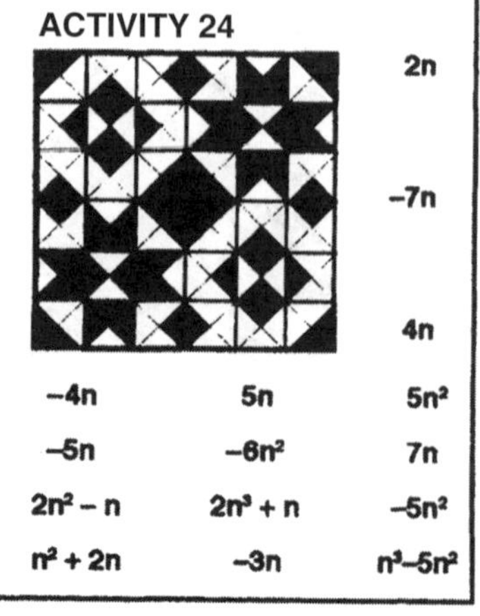

$2n$, $-7n$, $4n$

$-4n$	$5n$	$5n^2$
$-5n$	$-6n^2$	$7n$
$2n^2 - n$	$2n^3 + n$	$-5n^2$
$n^2 + 2n$	$-3n$	$n^3 - 5n^2$

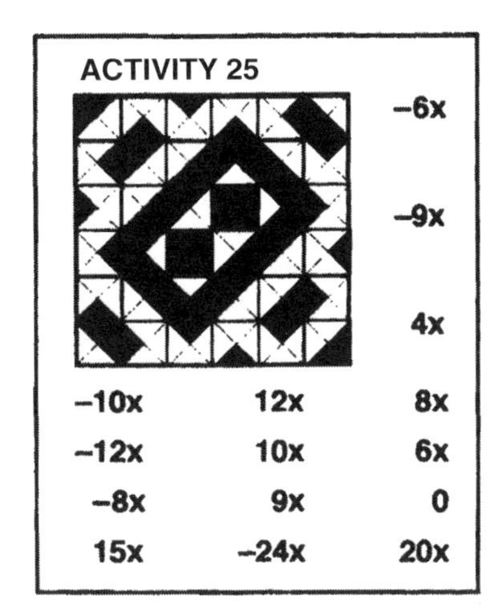
ACTIVITY 25
$-6x$
$-9x$
$4x$
$-10x$ $12x$ $8x$
$-12x$ $10x$ $6x$
$-8x$ $9x$ 0
$15x$ $-24x$ $20x$

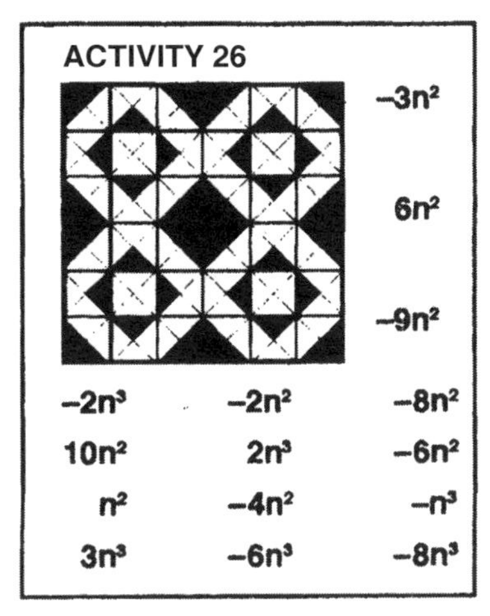
ACTIVITY 26
$-3n^2$
$6n^2$
$-9n^2$
$-2n^3$ $-2n^2$ $-8n^2$
$10n^2$ $2n^3$ $-6n^2$
n^2 $-4n^2$ $-n^3$
$3n^3$ $-6n^3$ $-8n^3$

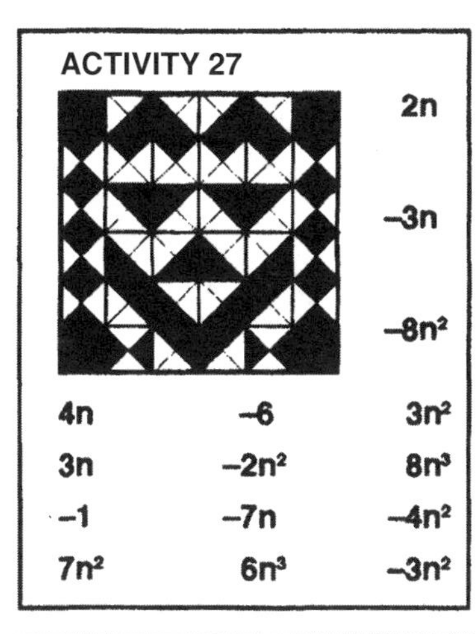
ACTIVITY 27
$2n$
$-3n$
$-8n^2$
$4n$ -6 $3n^2$
$3n$ $-2n^2$ $8n^3$
-1 $-7n$ $-4n^2$
$7n^2$ $6n^3$ $-3n^2$

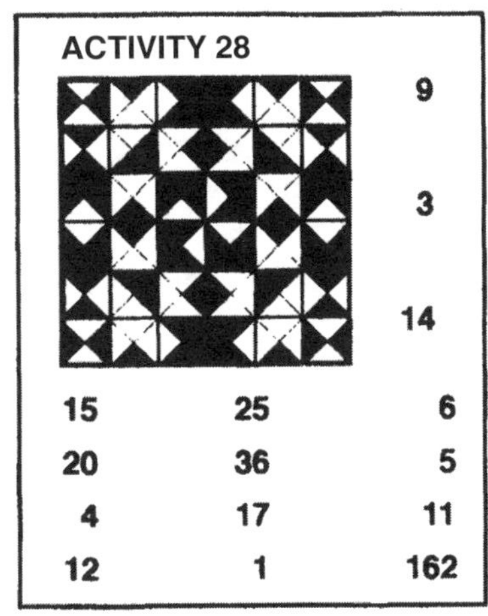
ACTIVITY 28
9
3
14
15 25 6
20 36 5
4 17 11
12 1 162

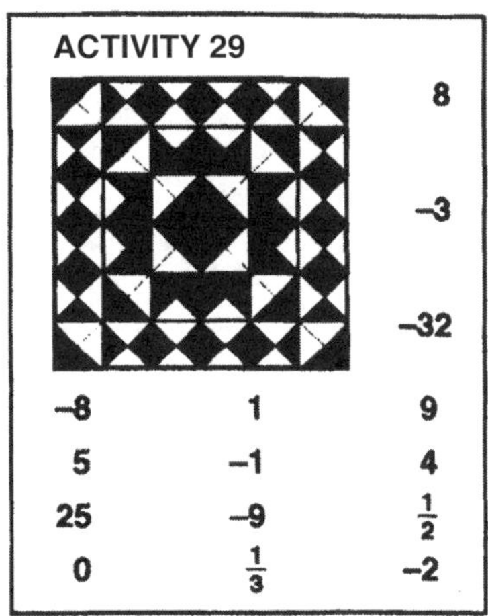
ACTIVITY 29
8
-3
-32
-8 1 9
5 -1 4
25 -9 $\frac{1}{2}$
0 $\frac{1}{3}$ -2

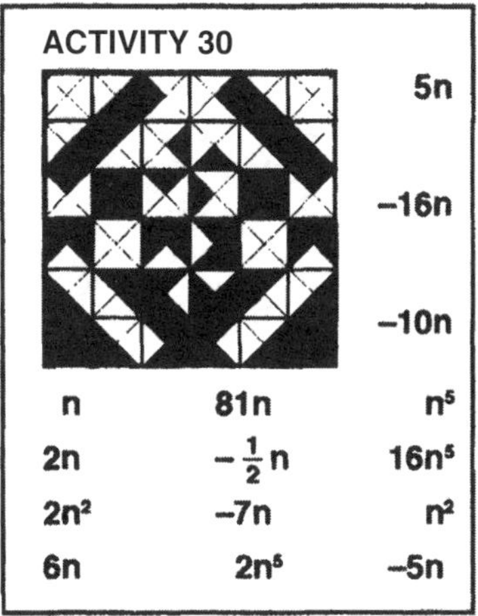
ACTIVITY 30
$5n$
$-16n$
$-10n$
n $81n$ n^5
$2n$ $-\frac{1}{2}n$ $16n^5$
$2n^2$ $-7n$ n^2
$6n$ $2n^5$ $-5n$

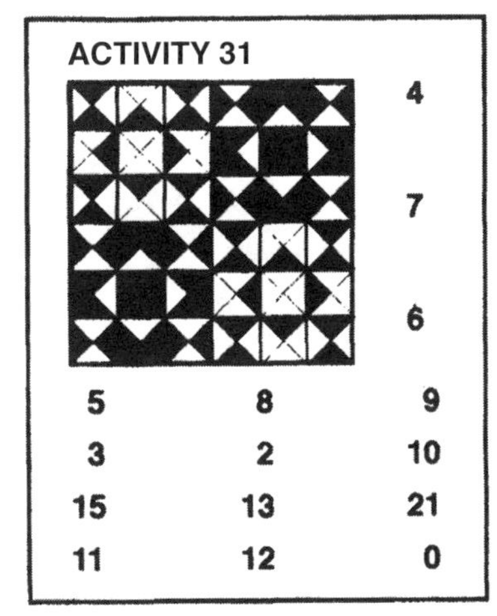
ACTIVITY 31
4
7
6
5 8 9
3 2 10
15 13 21
11 12 0

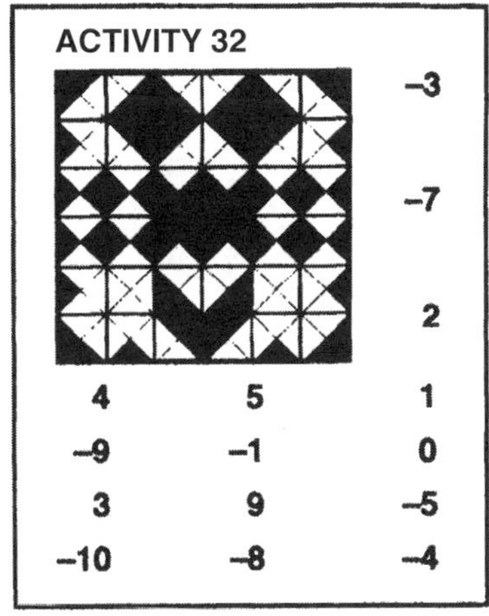
ACTIVITY 32
-3
-7
2
4 5 1
-9 -1 0
3 9 -5
-10 -8 -4

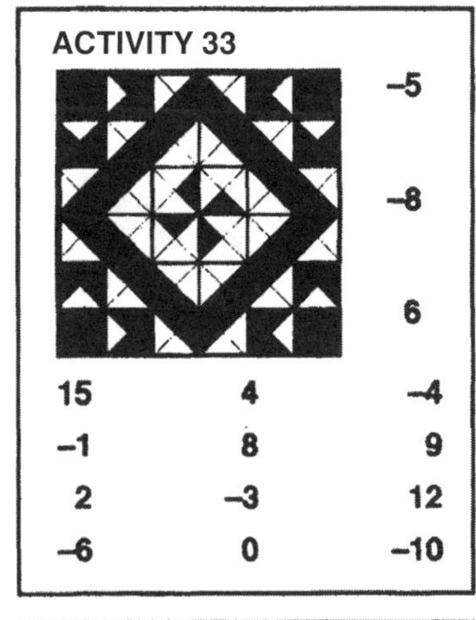
ACTIVITY 33
-5
-8
6
15 4 -4
-1 8 9
2 -3 12
-6 0 -10

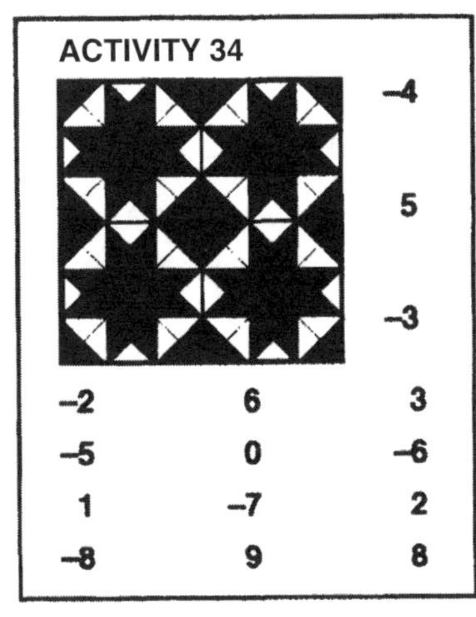
ACTIVITY 34
-4
5
-3
-2 6 3
-5 0 -6
1 -7 2
-8 9 8

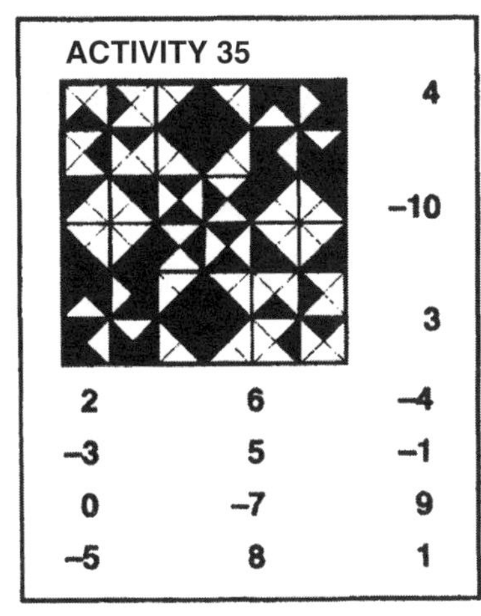
ACTIVITY 35
4
-10
3
2 6 -4
-3 5 -1
0 -7 9
-5 8 1

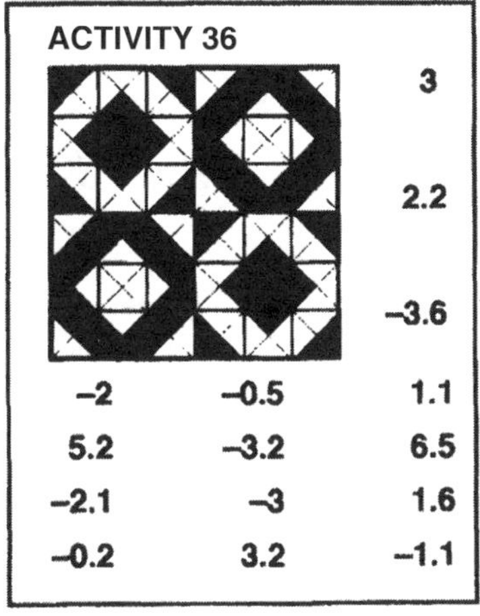
ACTIVITY 36
3
2.2
-3.6
-2 -0.5 1.1
5.2 -3.2 6.5
-2.1 -3 1.6
-0.2 3.2 -1.1

ACTIVITY 37

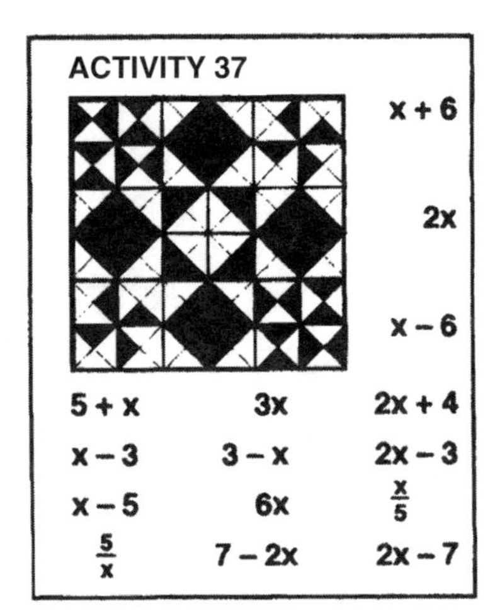

$x + 6$

$2x$

$x - 6$

$5 + x$	$3x$	$2x + 4$
$x - 3$	$3 - x$	$2x - 3$
$x - 5$	$6x$	$\frac{x}{5}$
$\frac{5}{x}$	$7 - 2x$	$2x - 7$

ACTIVITY 38

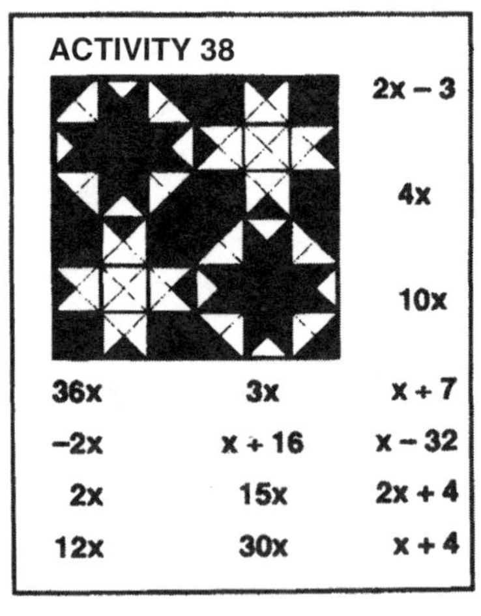

$2x - 3$

$4x$

$10x$

$36x$	$3x$	$x + 7$
$-2x$	$x + 16$	$x - 32$
$2x$	$15x$	$2x + 4$
$12x$	$30x$	$x + 4$

ACTIVITY 39

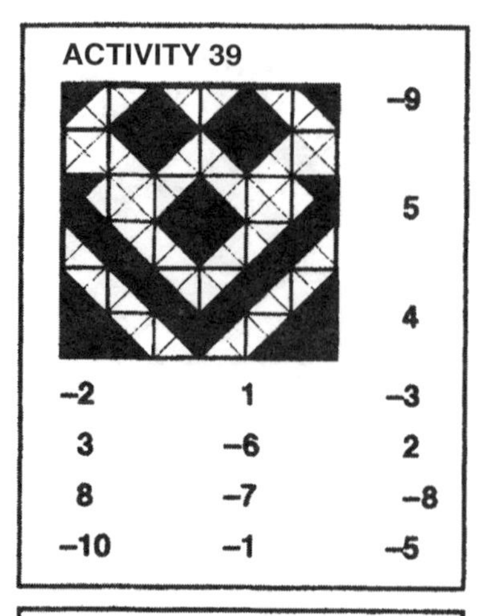

-9

5

4

-2	1	-3
3	-6	2
8	-7	-8
-10	-1	-5

ACTIVITY 40

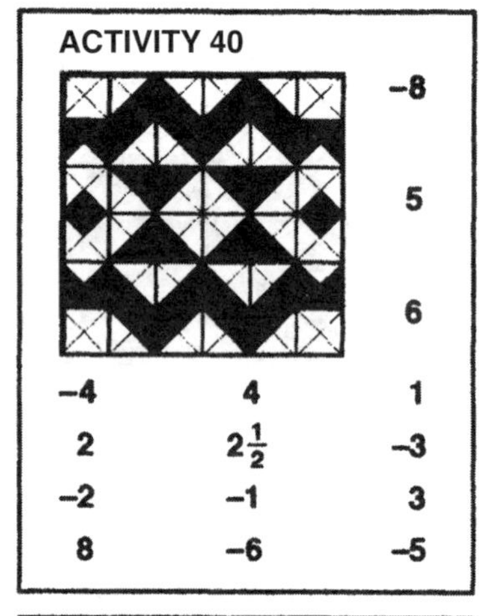

-8

5

6

-4	4	1
2	$2\frac{1}{2}$	-3
-2	-1	3
8	-6	-5

ACTIVITY 41

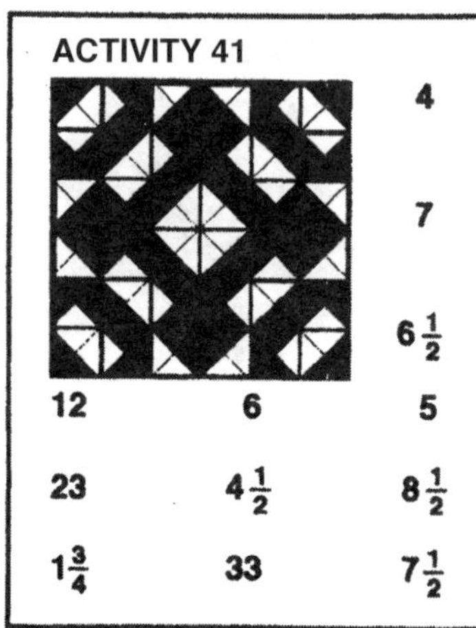

4

7

$6\frac{1}{2}$

12	6	5
23	$4\frac{1}{2}$	$8\frac{1}{2}$
$1\frac{3}{4}$	33	$7\frac{1}{2}$

ACTIVITY 42

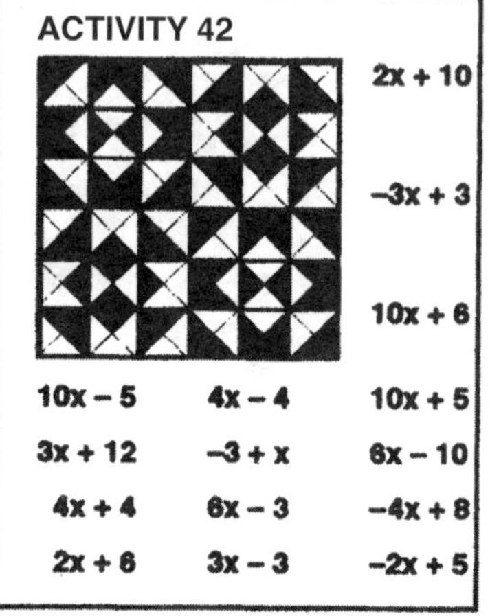

$2x + 10$

$-3x + 3$

$10x + 6$

$10x - 5$	$4x - 4$	$10x + 5$
$3x + 12$	$-3 + x$	$6x - 10$
$4x + 4$	$6x - 3$	$-4x + 8$
$2x + 6$	$3x - 3$	$-2x + 5$

ACTIVITY 43

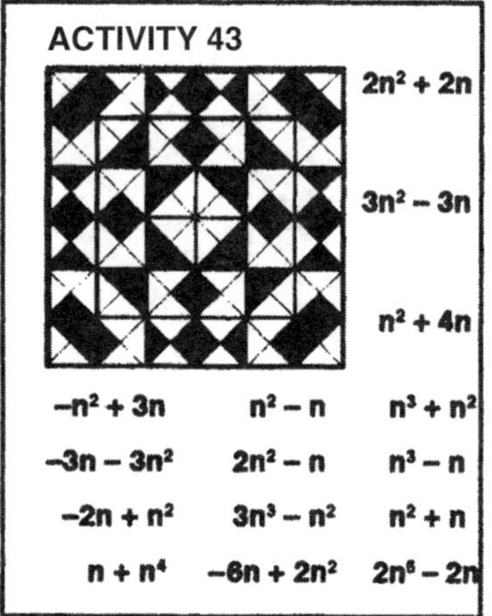

$2n^2 + 2n$

$3n^2 - 3n$

$n^2 + 4n$

$-n^2 + 3n$	$n^2 - n$	$n^3 + n^2$
$-3n - 3n^2$	$2n^2 - n$	$n^3 - n$
$-2n + n^2$	$3n^3 - n^2$	$n^2 + n$
$n + n^4$	$-6n + 2n^2$	$2n^6 - 2n$

ACTIVITY 44

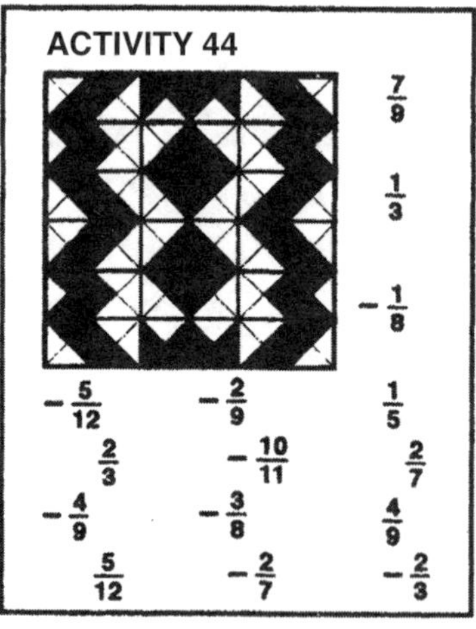

$\frac{7}{9}$

$\frac{1}{3}$

$-\frac{1}{8}$

$-\frac{5}{12}$	$-\frac{2}{9}$	$\frac{1}{5}$
$\frac{2}{3}$	$-\frac{10}{11}$	$\frac{2}{7}$
$-\frac{4}{9}$	$-\frac{3}{8}$	$\frac{4}{9}$
$\frac{5}{12}$	$-\frac{2}{7}$	$-\frac{2}{3}$

ACTIVITY 45

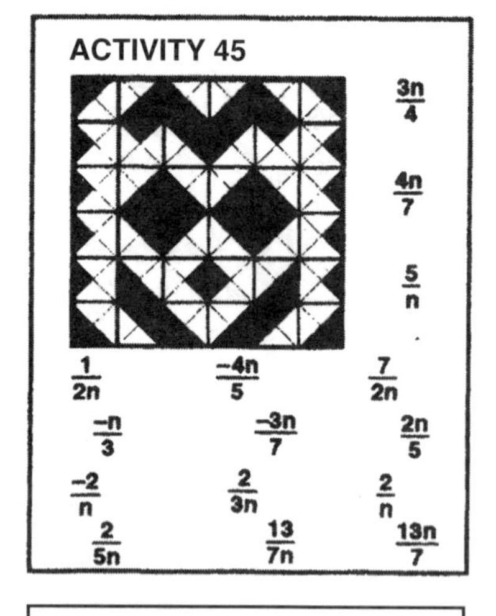

$\frac{3n}{4}$

$\frac{4n}{7}$

$\frac{5}{n}$

$\frac{1}{2n}$	$\frac{-4n}{5}$	$\frac{7}{2n}$
$\frac{-n}{3}$	$\frac{-3n}{7}$	$\frac{2n}{5}$
$\frac{-2}{n}$	$\frac{2}{3n}$	$\frac{2}{n}$
$\frac{2}{5n}$	$\frac{13}{7n}$	$\frac{13n}{7}$

ACTIVITY 46

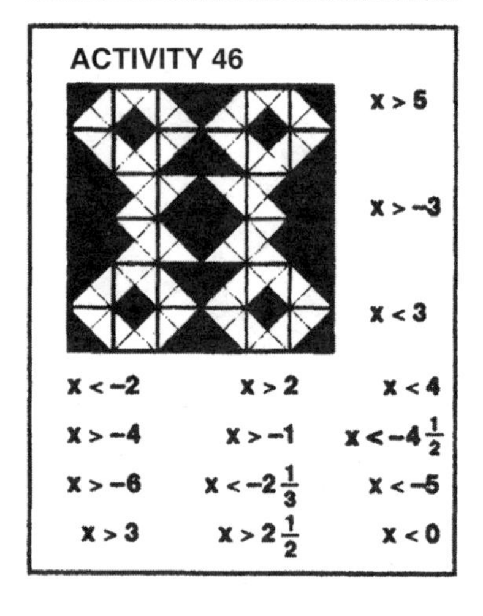

$x > 5$

$x > -3$

$x < 3$

$x < -2$	$x > 2$	$x < 4$
$x > -4$	$x > -1$	$x < -4\frac{1}{2}$
$x > -6$	$x < -2\frac{1}{3}$	$x < -5$
$x > 3$	$x > 2\frac{1}{2}$	$x < 0$

ACTIVITY 47

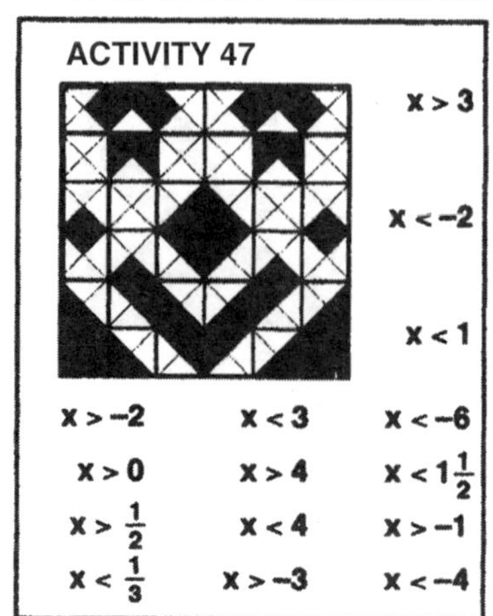

$x > 3$

$x < -2$

$x < 1$

$x > -2$	$x < 3$	$x < -6$
$x > 0$	$x > 4$	$x < 1\frac{1}{2}$
$x > \frac{1}{2}$	$x < 4$	$x > -1$
$x < \frac{1}{3}$	$x > -3$	$x < -4$

ACTIVITY 48

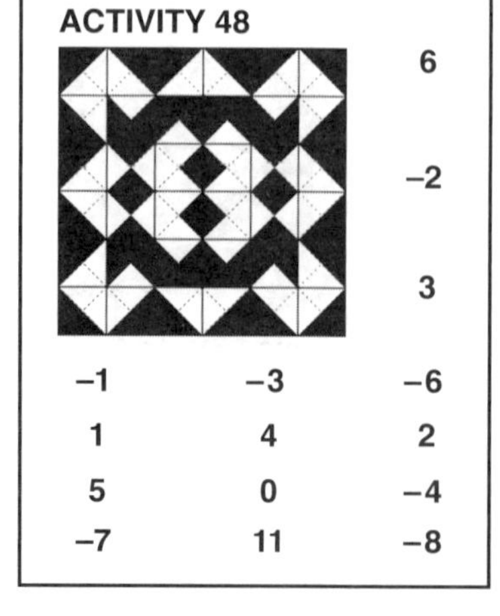

6

-2

3

-1	-3	-6
1	4	2
5	0	-4
-7	11	-8